NEUROSCIENCE:
A LABORATORY MANUAL

JAMES E. SKINNER, Ph.D.

Neurophysiology Section
Physiology Department
Baylor College of Medicine
and
Neurophysiology Department
The Methodist Hospital

SAUNDERS BOOKS IN PSYCHOLOGY
Robert D. Singer, *Editor*

W. B. SAUNDERS COMPANY
PHILADELPHIA · LONDON · TORONTO

W. B. Saunders Company: West Washington Square
Philadelphia, Pa. 19105

12 Dyott Street
London, WC1A 1DB

1835 Yonge Street
Toronto 7, Ontario

Neuroscience: A Laboratory Manual   SBN 0-7216-8345-2

Print No: 9   8   7   6   5   4   3   2

To Professor Donald B. Lindsley,
who taught me everything I know,
and to Professor Peter Kellaway,
who gave me everything I have, including a typewriter,
but most of all
to Thelma Jean,
who adds sparkle to everything I know and have.

FOREWORD

James E. Skinner, who has conceived and assembled this book, is an amazing young man. One might say he is a rare phenomenon. He is intuitive, impulsive, and innovative. His enthusiasm and energy are boundless. His determination and dedication are impressive. His insight and conclusions are sometimes astounding.

Initially he considered producing only a dissection guide for the cow or sheep brain and a stereotaxic atlas for the rat brain. He wanted to outline some of the steps he had taken and some of the experiences he had in learning something about the structure and function of the nervous system as a foundation for understanding psychological processes and behavior. One thing led to another as he began to retrace his own course of development; he added stereotaxic surgery, implantation of electrodes for stimulating and recording, and a bit on histological examination of the brain. Then came more and more details about method and procedure: how to make gross and microelectrodes, cryoprobes and the like; how to do this and do that. Next came some functional neuroanatomy in order to relate knowledge of brain structure to concepts of psychology and behavior. Finally, since electrophysiology has played such a prominent role in the building of bridges between brain and behavior during the past forty years, he found it necessary to add Chapter 1, providing a look at some fundamental electrophysiological concepts. Since electrophysiology is the main methodological approach he describes, it seemed desirable to add a little about electricity and electronics as these apply to physiological recording and stimulating systems.

Thus the book has grown. I believe it can serve a real and valuable purpose in introducing the beginning student to brain and behavior study, whether he be destined to become a physiological psychologist, a neurosciences student, a neurologist, or any other person with interest in brain and behavior relationships. Even the more advanced student can find it a valuable reference source, particularly with respect to the specialized techniques described.

Obviously it is not a comprehensive book in this field. If it were, the text materials would have to be several times increased. There are good reasons for keeping such a book brief and concise in what it attempts to cover. The research literature is so vast and the amount of information so great that any attempt at breadth and comprehensiveness is bound to fall short of its mark. But there is another reason for brevity and sinking a single shaft or two in the subject matter, even though this means, as is true in the case of this book, that it presents effectively only the essential sides of the picture methodologically and content-wise. The student will find that he can organize and cope with only so much information. It may not make too much difference just what that information is as long as it gains and holds his interest and gives him the impression that he is making some progress toward his goal. If he becomes vitally interested and motivated and finds that he can begin to use the information at once in his thinking and in even simple investigative steps, he will eventually get down to all of the bedrock fundamentals that he needs even though it may take him a lifetime to do so. Once he "catches on fire" nothing will hold him back. But, in the field of this book, if he had to take step by step all of the seemingly important prerequisites such as anatomy, neuro-anatomy, physiology, neurophysiology, biochemistry, and others before he could even pursue in the laboratory or in his head a simple investigative idea, he might be "turned off" before he ever got "turned on."

I believe Dr. Skinner's book, written in a style of communication that is informative, driving, and convincing will turn students on.

DONALD B. LINDSLEY

ACKNOWLEDGMENTS

I would like to acknowledge the assistance of the following students, colleagues and friends who contributed to the generation of the materials in this publication: Mara McDonald (stereotaxic atlas), Dennis Akutagawa (dissection guide), Sharon Bermeister (stereotaxic surgery), Harvey Michaels (photographs for drawings), Ben Benner and Larry Culpepper (simple cryogenic system), Sandy Strickler (mechanical drawings), and the many others who participated, including Jim Martin, Len Gardner, Connie Kovar, Walt Salinger, Steve Clarke and Cindy Wilson.

I wish to convey my sincere appreciation to the Scurlock Foundation of Houston, Texas, for their financial assistance. Without their backing, the high quality of the artwork could not have been achieved, and I would like to thank personally Mr. J. S. Blanton of the Scurlock Foundation, and Mr. Ted Bowen, Chief Administrator of The Methodist Hospital, for making the arrangements for the support.

CONTENTS

CHAPTER 1

ELECTRICAL ACTIVITY OF THE BRAIN 1

Information Processing 1
The Membrane of the Neuron 6
Postsynaptic Potentials.. 8
Electrogenesis of EEG Potentials...................................... 14
Recording Bioelectric Potentials 17
References 22

CHAPTER 2

FUNCTIONAL NEUROANATOMY AND DISSECTION GUIDE FOR THE COW AND SHEEP BRAINS 25

Specific Systems 27
Association Systems 29
Nonspecific Systems 31
Emotion Systems.. 34
Dissection Guide 39
References 83

CHAPTER 3

STEREOTAXIC SURGERY AND HISTOLOGICAL EXAMINATION OF THE BRAIN 87

Stereotaxic Surgery.. 88
Histological Examination of the Brain 120
Reference 143

CHAPTER 4

BASIC ELECTRONICS 145

Ohm's Law 145
Capacitance 150
Fourier Analysis of Wave Forms 153
Frequency Response... 153
Input Resistance 158
Common-Mode Rejection... 160
Gain and Noise.. 161

CHAPTER 5

CONSTRUCTION OF INTRACRANIAL IMPLANT DEVICES 163

Stimulation and Recording 163
Functional Blockade 169
Chemostimulation 178
Procedures for Construction of Various Implant Devices 179
List of Distributors for Materials 192
References 193

CHAPTER 6

STEREOTAXIC ATLAS OF THE RAT BRAIN 195

Subjects and Methods 195
Nomenclature 197
Stereotaxic Instrument Reference Points 197
Bregma-Midline-Cortical Surface Reference Points 201
Blocking Plane 202
Summary 202
References 204

**LIST OF ABBREVIATIONS FOR RAT ATLAS
INTERNATIONAL NOMENCLATURE OF BRAIN STRUCTURES** 205

**INDEX OF ANATOMICAL STRUCTURES OF THE COW AND SHEEP
BRAINS** 239

SUBJECT INDEX 241

CHAPTER 1

ELECTRICAL ACTIVITY OF THE BRAIN

INFORMATION PROCESSING

The early investigators of brain function visualized "energies" flowing from one part of the brain to another as the underlying mechanism for all higher nervous activities (Müller, 1838; Sherrington, 1906). The information-processing theories developed in the late 1940's (Shannon, 1949; Wiener, 1948) have made it apparent to biologists that it is *information* that flows between regions of the brain, and energy serves only as its carrier. Thus, each of several different carriers of energy converging upon a nerve cell contains compatible information for the cell to integrate while performing its higher nervous activities.

One of the important ways nerve cells are known to communicate with one another is through all-or-none, binary-encoded information transmitted by an electrochemical carrier called an *action potential*. During a special condition of excitability in a neuron, an action potential is generated from a trigger zone somewhere in the cell and is then propagated from this region down a long fibrillar structure, called an axon, in order to affect, in turn, the excitability of the cell or cells to which it projects. The action potential is propagated along the axons at speeds up to 100 m per second. The axons may be either highly bifurcated and collateralized, projecting to and making contacts with cells in widespread regions of the brain, or they may be limited in their projections and make contacts with only one or a few cells (Scheibel and Scheibel, 1966) (compare the RF cell to the SP cell in Fig. 1-1). During embryological development, axons grow from one region of the brain to another with great *specificity* and, as demonstrated in mature amphibians, if cut they will regenerate and re-form their original functional connections (Sperry, 1944, 1965).

When an action potential reaches one of the terminal endings of an axon, it causes the release from this part of the cell of a chemical

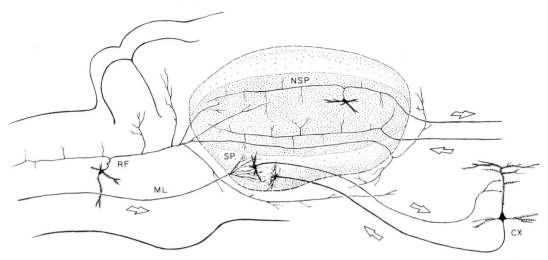

Figure 1–1 Golgi-stained cells of the specific and nonspecific thalamic systems. The Golgi method stains all the parts of a cell, but only a small percentage of the cells in the tissue. This enables one to make realistic drawings of the morphology of an entire cell (i.e., its dendrites, soma, and axon) and to examine it in all its details. Illustrated is the difference in the axon collateralization pattern of the cells in the specific (SP) and nonspecific (NSP) thalamic systems. The specific system, which serves as a relay of sensory information from the periphery into the higher brain centers, tends to have cells with noncollateralized axons which project to the dendritic fields of only one or two cells and end in bushy terminal arborizations. The nonspecific system, which apparently serves in complex higher nervous functions related to arousal, attention and learning, tends to have cells with highly collateralized axons which project to the dendritic fields of many different cells in many regions of the brain. The morphologic characteristics of these cells must surely be related to the different functions each system subserves. Note the spiny appearance of the dendrites on the cells. An electron micrograph of one dendritic spine is shown in Figure 1–2. As many as 10,000 synaptic contacts typically occur on each single cell. Many of these contacts are on the dendritic spines, but others are on the dendritic trunk, soma, axon hillock and axon terminals. Abbreviations: CX, pyramidal cell in the cerebral cortex; ML, axon in the medial lemniscus; NSP, axon from the cortex and cell in the nonspecific, midline nuclei of the thalamus; RF, cell in the brain stem reticular formation; SP, cell in the specific relay nuclei of the ventrobasal complex of the thalamus. (Drawing made after the work of Scheibel and Scheibel, 1967.)

substance stored in little packets or vesicles (Fig. 1-2, SV). Within 0.5 msec after its release (Fig. 1-3 A), the chemical produces a nonpropagated electrical response in the cell upon which it projects. This nonpropagated response is called a postsynaptic potential and, together with many others from other inputs, it affects the condition of excitability in the recipient cell. Each terminal ending, the space between it and its projection cell, and the part of the projection cell which responds to the chemical transmitter substance are all together called a *synapse*, and a typical one is shown in the electron micrograph in Figure 1-2, SP. A single cortical neuron may have as many as 10,000 synaptic inputs, each having individual bits of information that must be integrated at each moment in time in order to determine the cell's excitability or probability of producing an all-or-none action potential.

Electrochemical synaptic transmission is apparently not the only way neural cells can communicate with one another. Electron micrographs show fused *tight junctions* between the membranes of many

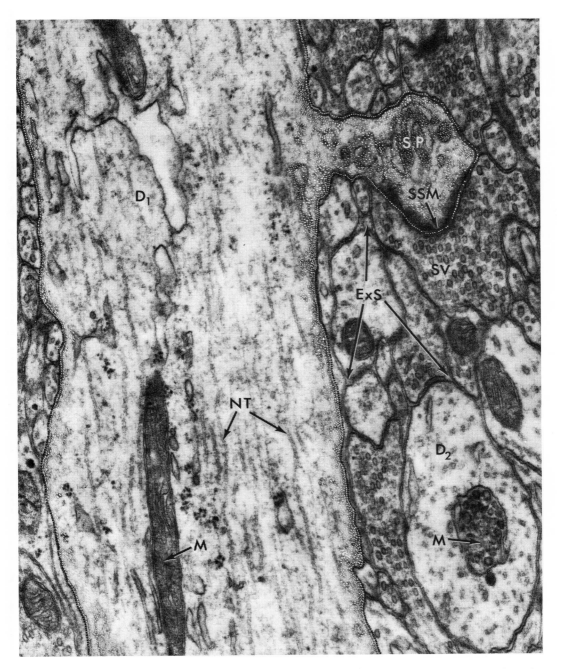

Figure 1–2 Electron micrograph of a cortical dendrite. This figure shows a longitudinal section of a dendrite (D_1) with an attached dendritic spine (SP). Note the thickening of the subsynaptic membrane (SSM) in the spine. Across the synaptic cleft from this specialized portion of the dendritic spine is a terminal ending of the presynaptic cell. This latter structure is filled with synaptic vesicles (SV) which are presumed to contain the transmitter substance of the synapse. A second, similar dendrite (D_2) is seen in cross-section in the lower right corner. Mitochondria (M) and neurotubules (NT) are associated with dendrites and help to identify their inclusive structures as dendrites rather than axons, cell bodies, and so forth. Note the tight packing of the structures, which exclude much of the extracellular space (ExS) between them. In the alive and intact brain this space between the cells has been estimated to be around 18 to 25 per cent of the total volume. In the preparation of tissue for examination with the electron microscope, this space is apparently reduced, probably by moving water from the extracellular space into the cells before their membranes are chemically fixed. (Electron micrograph courtesy of Dr. David Maxwell; $\times 35,000$).

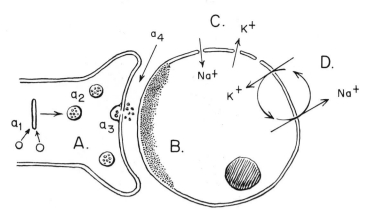

Figure 1-3 Various "energies" involved in the transmission of neural information. *A,* Transmitter substances. Chemical transmitters are synthesized (a₁), moved to their point of release within the presynaptic cell (a₂), released into the synaptic cleft in order to affect the postsynaptic membrane (a₃), and then hydrolyzed by enzymes back into precursor elements in order to neutralize the effect upon the postsynaptic membrane (a₄). *B,* Subsynaptic membrane. There is some evidence that the subsynaptic membrane may determine whether a given transmitter is hyperpolarizing or depolarizing in its effect upon the postsynaptic cell. *C,* Membrane changes during depolarization. It has been proposed that holes are opened in the membrane by a rapid change in the conformation of the membranal protein. *D,* Polarization of the membrane. Metabolic factors, as well as concentrations of extracellular ions, determine the degree to which the membrane can be polarized.

kinds of cells in the body, including nerve cells. Physiological studies of these cells show *low resistance junctions* between two adjacent cells in which the electrical activity in one can be transmitted through a low electrical resistance directly to the other without an intervening chemical transmitter (Furshpan and Potter, 1968). The role of this electrical coupling in regulating the excitability of a neuron or its information-processing mechanism is not yet known, but it is clear that this manner of information transmission between cells is potentially of great importance. It is the combined set of electrical, chemical and other unknown energies impinging upon a neuron that controls the electrical excitability and determines when an information-carrying action potential is to be transmitted.

Many kinds of living organisms are able to learn and remember how to react to stimuli in an ever-changing environment. It then follows that there must be some means for adaptively changing the simple integrative information-processing capabilities of the nervous system and some mechanism for storing this change. How and where information is acquired and stored in the brain has been the subject of investigation and controversy for many years and is by no means clearly understood at present. The degradation of a learned and remembered behavioral response by the removal of brain tissue in animals has revealed deficits which seem to be proportional to the *amount* of tissue removed and are independent, within limits, of the exact *place* of removal (Lashley, 1930, 1950; principles of *mass action* and *equipotentiality*). A single cell only seems to participate in a larger neural mechanism for learning and memory and by itself is not of crucial importance; but still some cells in this larger mechanism must contain the changeable information-processing mechanisms for learning and memory in addition to those passive ones which integrate the inputs and, without adaptive change, determine the outputs.

The creation of *reverberating neural circuits, anatomical growth* and changes in *biochemical compounds* have all been proposed as active mechanisms which could acquire and store information at the cellular level. Jamming or depressing the electrical circuits in the brain by electroconvulsive shock or cooling would be expected to disrupt or inactivate the many neural circuits hypothesized to be reverberating and to underlie memory; but this procedure does not seem to affect the later recall of well established memories in subjects so tested, though it does seem to abolish newly learned or short-term memories that have not yet been consolidated by the brain into long-term memories (McGaugh, 1965).

Anatomical growth has been reported for both synaptic terminal endings (Szentagothai and Rajkovits, 1955) and large regions of the sensory cortex (Krech et al., 1963) after these tissues have been repeatedly exposed to stimulation through their peripheral sensory pathways by increased or enriched environmental stimuli. Brain tissue either does not develop properly (Wiesel and Hubel, 1963) or atrophies and disappears (Cajal, 1911, 1928; Reisen, 1961) in a sensory system that is not exposed to stimulation. Whether the hypertrophy of use and atrophy of disuse can play a role in memory storage and forgetting remains to be proved, especially for those quickly learned memories which occur so rapidly that anatomic growth seems impossible.

More recently, experiments in the biochemical transfer of memory have hinted at a third mechanism for neural storage (Ungar, 1970). In these experiments a chemical brain extract collected from a trained animal is injected into an untrained animal which, after injection, exhibits the learned behavior. It is clear that such brain extractions, each containing a mixture of RNA, polypeptides and unidentified biological compounds, are able to transfer something between the animals, but it has not yet been clearly demonstrated whether a memory compound was transferred or merely a sensitizing agent, such as a hormone, which caused the release of a response pattern that was already present in the animal's behavioral repertory but was subthreshold until the releasing factor was present.

Nerve cells are not the only structural elements within the brain. They are surrounded by an extracellular space which is filled with a gluelike substance containing many different biochemical compounds, and by glial cells which contain and transfer many metabolic and biochemical substances to the neurons. Several theories of learning and memory are based on glial cells and emphasize their transfer of RNA to the neurons (Hydén, 1959; Galambos and Morgan, 1960; Robinson, 1966; John, 1967). Most of the supporting evidence for these theories is, however, only suggestive and is not as yet conclusive.

Weak currents generated by each individual neuron in the brain travel in the extracellular space (Elul, 1968) and perhaps influence the excitability of other neurons they pass. Electrical currents produced by electrodes, and of the same order of magnitude as those available in the extracellular space, can affect the electrical excitability of some neurons; and the type of effect, either inhibitory or excitatory, depends upon the direction of flow of the induced currents (Terzuolo and Bullock, 1956; Calvet et al., 1964). Adey et al (1966) have discussed how impedance changes may reflect structural changes of the extracellular space, and they have shown rapid impedance changes in the brain during certain learned behavioral tasks. They have also shown how certain ions such as calcium are able to maintain these impedance changes and perhaps play an important role in information storage in the brain (Wang et al., 1966).

The integrative and adaptive information-processing mechanisms of the brain most likely utilize a combination of some or all of the anatomic, biochemical, electrical and other as yet unknown "energies" which carry information to and away from nerve cells and participate in their higher nervous mechanisms. Even the life-sustaining metabolic mechanisms thought to be unimportant in information processing do have an influence on the excitability and transmission properties of neurons, and under certain conditions do affect the higher nervous functions. Whatever the various modes of interaction between nerve cells might be, it is traditionally believed that these will either directly or indirectly affect the electrical activity of the brain.

THE MEMBRANE OF THE NEURON

The regenerative mechanism for propagating the information-carrying action potential along the axon of a nerve cell depends upon two submechanisms: one for maintaining a potential energy source, known as the *transmembrane potential,* which appears between the inside and outside of a cell's membrane (Fig. 1−3 *D*), and another for opening small *current channels* in the membrane through which the membrane potential can discharge (Fig. 1-3 *C*). The action potential is produced by the discharge of the membrane potential through a momentary low resistance pathway which regeneratively moves or propagates along the axon.

There are two convincing theories regarding the formation of the membrane potential and its discharging mechanism; in both the membrane potential is believed to be produced by a greater number of sodium ions and a smaller number of potassium ions on the outside of the cell than on the inside. These combined concentration differences produce a net potential difference across the membrane, with the negative pole being on the inside. Hodgkin and Huxley (see Hodgkin, 1964) first proposed that the *active transport* of these ions by a "sodium-potassium pump" keeps sodium out and potassium inside the cell, and that a *permeability change in the membrane* (e.g., pores open in the membrane) allows the ions to flow back through it under the pressure of their concentration gradients and discharge the membrane potential. Ling (1964, 1965, 1969) has been unable to account for the large amount of energy required to operate such a hypothesized sodium-potassium pump and has proposed an alternative mechanism for establishing the different ionic distributions across the membrane. Sodium ions have several states of hydration with free water, and if the water is bound with other molecules, as it appears to be inside the cell, then the ions will move outside the cell where they have *greater freedom for hydration* (law of entropy). By electrically or chemically changing one part of the complex molecular structure of the membrane and cytoplasm of the cell, an *induced change of the binding sites* in other parts of the structure occurs which makes it more favorable for sodium to bind and hydrate inside the cell. This causes the ions to move under their own entropic pressures back across the membrane to discharge the membrane potential. Ling's theoretical mechanism explains the same phenomena as the one proposed by Hodgkin and Huxley, but operates on an amount of energy known to be available in the brain tissue, in contrast to that of Hodgkin and Huxley, which seems to require a great deal more.

When an action potential is initiated in the neuronal membrane, current flows into the cell at that particular point through a low resistance channel, and an equal amount of current flows out through other, higher resistance regions of the cell. If there are no insulators on the membrane, most of the outward current will flow through either side of the initial action potential's point of inward current flow. This outward flow causes a conductance change in the membrane i.e., lower resistance) which allows the membrane potential to again discharge an inward current and initiate successive action potentials at these more lateral points. An action potential over one point on the membrane can sustain its inward flow of current only for a brief time before it stops and becomes refractory to further discharge. If the outward current on one side of a new action potential flows through a refractory piece of membrane which previously supported an action potential, then it will be ineffective in producing a successive action potential at this old location, while on the other side of the new action potential the outward current which flows through fresh or rejuvenated membrane will be successful. This *refractory phenomenon* causes each successive action potential to be propagated away from its original point of initiation and from each preceding one in the regenerative sequence.

If a high resistance insulator is wrapped around the axon at a point adjacent to an action potential, the outward current cannot flow through the membrane at this point but must seek a more distant, uninsulated region. The myelin sheath wrapped around the axons of many neurons in the brain increases their axonal propagation velocity with this insulation by causing the outward currents to reach out to the uninsulated regions, called nodes of Ranvier (Fig. 1–4 *B*, NR), and creates successive action potentials there which skip or jump past the insulated regions;

this phenomenon is known as *saltatory conduction*, which in Latin means "skipping" conduction.

The mechanism for the discharge of the membrane potential is not yet known, but it is clear that both electrical currents and several types of chemical compounds are effective in regulating the electrical excitability of the nerve cells. These various currents and compounds have both excitatory and inhibitory effects. The injection of a weak transmembranal electrical current through an intracellular microelectrode will either produce or inhibit the production of an action potential in the neuron (Eccles, 1964), and small amounts of certain chemical compounds injected near the extracellular membrane of a neuron will increase or decrease its electrical discharges (Salmoiraghi and Stefanis, 1967). Dale's work (1935, 1937) has demonstrated that one individual nerve cell secretes only one kind of chemical substance, and it is commonly believed today that Dale's hypothesis can be extended to cover the neurotransmitter substances. Kandel et al. (1966) have shown that one type of transmitter substance (acetylcholine) secreted from a single neuron can have both an excitatory effect on one cell and an inhibitory effect on another of the cells to which its bifurcated axon projects. Thus, a given neurotransmitter can be either excitatory or inhibitory, depending upon the properties of the *postsynaptic membrane* (Fig. 1-3 B). Whatever the source (electrical or chemical) or the sign (inhibitory or excitatory) of each of these presynaptic inputs, the result will be a postsynaptic effect which in most cases will produce an electrical response that can be recorded across the cell membrane.

POSTSYNAPTIC POTENTIALS

SPINAL CORD

When electrical currents or particular chemical substances are applied to the membrane of a neuron, a nonpropagated *local response* is recorded from an electrode placed inside the cell. A local response is like an action potential, but its outward return currents through the membrane are not strong enough to produce successive responses. The potential does not regenerate and propagate but remains locally on the membrane where it is initiated. Two types of local responses can be produced; those which tend to *depolarize* the membrane potential and send outward transmembranal return currents, like the action potentials; and those which tend to *hyper-polarize* the membrane potential and send return currents in the opposite direction.

To record the local potentials, the tip of a small glass microelectrode is placed inside the neuron in order to record between the inside and outside of the membrane. Microelectrodes are thought to be too large to fit inside any part of a cell other than the bulging cell body without damaging the membrane and impairing the function and electrical properties of the cell. The cells chosen for recording are usually the large motoneurons in the ventral root of the spinal cord (VR cells) which in-

nervate the limb muscles. The VR cell receives inputs from nerves attached to sensory receptors, located in the muscles, that detect muscle
stretch. Each sensory nerve is from a muscle that is either synergistic or
antagonistic to the muscle that the VR cell causes to contract. Stimulation either directly by electricity or naturally by muscle stretch produces
either a depolarization or a hyperpolarization shift in the recorded
potential, depending upon which particular nerve is activated. With no
input to the VR cell, the microelectrode records only the steady, negative
transmembrane potential or resting potential (E_R in Fig. 1–4 C).

When the synergistic (excitatory) nerve is briefly stimulated, an
excitatory postsynaptic potential (EPSP) is produced. The recorded
membrane potential becomes more positive or depolarized and, according to Eccles (1964), moves toward the equilibrium potential for the
EPSP (E_{EPSP} in Fig. 1-4 C). The EPSP is thought to result from the
increased permeability of the membrane to the large and small hydrated
ions, which tends to move the membrane potential toward complete
depolarization, with the sodium rushing into the cell and the potassium
rushing out.

When the antagonistic (inhibitory) nerve is stimulated, an *inhibitory postsynaptic potential* (IPSP) is produced, and the recorded potential inside the cell becomes more negative or hyperpolarized. The IPSP
is thought to result from the increased permeability of the membrane to
only the small hydrated ions (K^+ and Cl^-), which moves the membrane
potential toward the net equilibrium potential of these ions. When a
cathodal current is passed through a glass microelectrode filled with a
chloride salt, the anions move out of the microelectrode and into the
cell, thus increasing the intracellular concentration of Cl^- ions and
shifting the chloride equilibrium potential toward a more depolarized
level. If enough Cl^- ions are injected, the equilibrium potential for the
IPSP (E_{IPSP} in Fig. 1-4 C) will become more positive than the resting
potential, and an evoked IPSP will move in a depolarized direction
rather than in the normal hyperpolarized direction. This technique is
often used to confirm that a recorded postsynaptic hyperpolarization is
indeed an IPSP.

EPSP's show spatial and temporal summation. If two different excitatory nerves of synergistic muscles are stimulated at the same time,
their resultant EPSP's in the single VR cell will be additive *(spatial
summation)* and will produce a larger depolarization of the membrane
potential than would occur if only one nerve were stimulated. If one of
these nerves is stimulated several times in rapid succession so that the
EPSP's overlap in time, they too will be additive *(temporal summation)* and similarly produce a larger depolarization. The important consequence of this spatial and temporal summation is that the membrane
potential can be depolarized enough to reach a critical firing level (E_{CFL}
in Fig. 1-4 C), after which regenerative depolarization begins to occur
and produce a propagated action potential. In some experiments it has
been shown that IPSP's superimposed upon summating EPSP's are able
to swing the membrane potential from depolarization back into hyperpo-

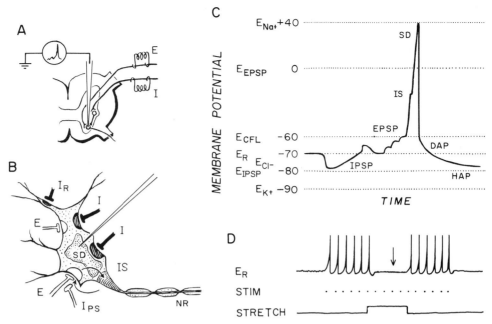

Figure 1–4   Intracellular records from spinal cord neurons. *A,* Illustration of the preparation: E, excitatory stimulus; I, inhibitory stimulus. *B,* Diagrammatic representation of a spinal cord neuron impaled by a microelectrode: E, excitatory synapse; I, inhibitory synapse; I_R, "remote" inhibitory synapse; I_{PS}, "presynaptic" inhibitory synapse; IS, initial segment trigger zone; SD, somatic and dendritic membrane capable of regenerative depolarization; NR, node of Ranvier. *C,* Diagrammatic record showing various intracellular potentials: IPSP, inhibitory postsynaptic potential; EPSP, excitatory postsynaptic potential; IS, initial segment discharge; SD, soma-dendritic discharge; DAP, depolarizing afterpotential; HAP, hyperpolarizing afterpotential. The various membrane and ionic equilibrium potentials are shown to the left with their respective voltages (in millivolts): E_{EPSP}, equilibrium potential for EPSP's; E_{CFL}, critical firing level; E_R, resting membrane potential; E_{IPSP}, equilibrium potential for IPSP's; E_{Na}^+, equilibrium potential for sodium ions; E_{Cl}^-, equilibrium potential for chloride ions; E_K^+, equilibrium potential for potassium ions. *D,* Illustration of results implying "remote inhibition," from the work of Frank (1959) and Granit et al. (1964): E_R, resting membrane potential; STIM, postsynaptic transmembranal electrical stimuli at threshold for producing action potentials; STRETCH, stretch of antagonistic (inhibitory) muscle. Note that during STRETCH, the transmembranal electrical stimuli are no longer effective in producing action potentials and there is no recorded sign of hyperpolarization (arrow).

larization before the critical firing level is reached, with the important consequence that IPSP's can inhibit the excitatory effect of EPSP's that would otherwise produce a cell discharge.

The critical firing level (CFL) of a VR cell remains stable under optimal conditions of recording in the spinal cord, and once the depolarization of the membrane potential reaches this level, the cell always produces an action potential. The stability of the CFL is thought to be attributable to a patch of membrane located at the axon hillock or initial segment (IS in Fig. 1-4 *B*), whose threshold for regenerative depolarization is lower than that of any other part of the cell membrane. EPSP's cause current to flow into the cell at their site of generation, but an equal amount of return current must flow out of the cell at the same time, part of it flowing out through the IS. When EPSP's are numerous enough to cause a critical amount of outward current to flow through the IS, an action potential is produced. When the lower threshold IS discharges, it triggers an action potential down the axon and also sends an action potential in the opposite direction back into the soma and dendrites of the cell. The microelectrode records first the strong regenerative depolarization response of the initial segment (IS in Fig. 1-4 C) and then the depolarization of the soma and dendrites (SD in Fig. 1-4 C) as the action potential sweeps back over them. The axonal action potential is, of course, propagated in the normal orthodromic direction away from the initial segment toward its terminal endings.

When an action potential occurs, there are two different ionic currents which flow (Hodgkin and Huxley, 1952; performed on invertebrate axons). First, Na^+ flows into the cell, which is thought to shift the membrane potential toward the equilibrium potential for sodium ions, E_{Na+}, and then, 0.5 msec later, K^+ ions begin to flow out of the cell and the flow of Na^+ ions begins to stop, which is thought to shift the membrane potential in a negative direction toward the equilibrium potential for potassium ions, E_{K+}. The net result is the appearance of a depolarized afterpotential (DAP in Fig. 1-4 C) caused by the predominance of Na^+ ions flowing into the cell, followed by a hyperpolarized afterpotential (HAP in Fig. 1-4 C) caused by the predominance of K^+ ions flowing out of the cell. As might be expected, the excitability of the cell is greater during the DAP and less during the HAP (Gasser, 1939), since the membrane potential is nearer or farther, respectively, from the CFL.

Frank and Fuortes (1957) observed that the amplitude of an EPSP was diminished by as much as 50 per cent when preceded by electrical shocks to an antagonistic nerve, but there was no measurable hyperpolarization in the VR cell. Frank (1959) proposed that this reduction in the amplitude of the EPSP was caused by "remote inhibition," in which an IPSP was produced at a remote location on the membrane, e.g., the dendrites, but was not detected by the microelectrode, which was located far away in the soma where the currents of the postsynaptic potential did not flow (see Fig. 1-4 B, I_R).

In an organized series of experiments, Eccles and his associates (1964) established that the Frank and Fuortes observation could be explained by "presynaptic inhibition," in which stimulation of the inhibitory afferent nerve fibers caused a slight depolarization in the presynaptic terminal of the excitatory synapse which, in turn, made it less effective in generating its usual EPSP in the postsynaptic cell (I_{PS} in Fig. 1-4 B; note that the effect is on the presynaptic side of the synapse rather than on the postsynaptic side). The concept of remote inhibition was then abandoned because of the lack of definitive supporting evidence.

Later Granit et al. (1964) revived the earlier concept by performing an experiment that revealed inhibitory effects on VR cells but excluded the possibility for "presynaptic inhibition." They repetitively injected brief depolarizing currents into the cell through a microelectrode and adjusted the intensity for producing action potentials to the minimum threshold; that is, each stimulus pulse depolarized the membrane potential just to the CFL. Once the cell was firing regularly to each injected stimulus, the antagonistic (inhibitory) muscle was stretched, which caused the VR cell to stop firing, but as in the experiment of Frank and Fuortes, there was *no measurable hyperpolarization shift* in the membrane potential (see Fig. 1–4 D). A purely presynaptic inhibitory mechanism could not account for this inhibitory change in the postsynaptic membrane because there was no apparent presynaptic element involved in exciting the cell. The important point established by these studies on remote inhibition is that the excitability of a given VR cell is not always related to the postsynaptic potentials *that can be recorded* under optimal conditions. However, such a relation can be found often and in many cells in the spinal cord, as Eccles and his associates have shown.

CENTRAL NERVOUS SYSTEM

Large spontaneous fluctuations in the membrane potential in both depolarizing and hyperpolarizing directions have been recorded from many kinds of cells in the central nervous system, including those in the neocortex (Creutzfeldt et al., 1964; Elul, 1964; Jasper and Stefanis, 1965), hippocampus (Fujita and Sato, 1964) and reticular formation (Segundo et al., 1967). In many of these cells there does not seem to be a correlation between the membrane fluctuations and the cell's excitability, as determined by whether an action potential occurs. Purpura et al. (1965) have recorded intracellular potentials from immature cortex and have observed action potentials arising from hyperpolarizations (see Fig. 1-5 C, first spike on left). Elul and Adey (1966) observed spikes in the same cell which began from different CFL's (see Fig. 1-5 C, dashed lines). In the reticular formation, Segundo et al. (1967) recorded spike action potentials starting abruptly from the resting membrane potential or with no consistent relation to the recorded depolarizing potentials that preceded the spike discharge. Elul and Adey (1966) interpreted their results as demonstrating "remote excitation"; that is, trigger zones somewhere in the cell are producing action potentials, but the membrane fluctuations associated with the initiation of these action potentials are not recorded because they are remote and do not send current into the region of the recording electrode.

Evidence has been accumulating which indicates that some cells in the central nervous system, perhaps unlike VR spinal-cord neurons, may have low threshold action-potential *trigger zones located in the dendrites* as well as at the initial segment. In hippocampal pyramidal cells, Spencer and Kandel (1961) were able to observe two types of potentials preceding the spike discharge, a fast prepotential (FPP) as well as an IS inflection. This pattern occurred only when the cell was activated by a normal *orthodromic* stimulus arriving through synaptic contacts with its presynaptic cells (see Fig. 1-5 D). However, when the recorded cell was fired by an *antidromic* action potential initiated in its axon and sent backward into its soma and dendrites, only the IS inflection was ob-

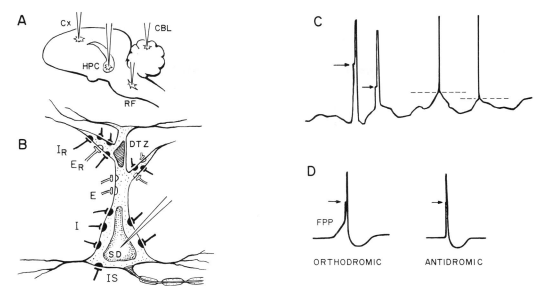

Figure 1–5 Intracellular records from central nervous system neurons. *A,* Illustration of the preparation showing the locations where recordings have revealed complexities which distinguish these cells from those in the spinal cord: Cx, cerebral cortex; HPC, hippocampus; RF, reticular formation; CBL, cerebellum. *B,* Diagrammatic representation of a cortical neuron impaled with a microelectrode: E, excitatory synapse; I, inhibitory synapse; I_R, "remote" inhibitory synapse; E_R, "remote" excitatory synapse; IS, initial segment trigger zone; SD, somatic and dendritic membrane capable of regenerative depolarization; DTZ, dendritic trigger zone. *C,* Diagrammatic record showing various intracellular potentials. The slow-wave fluctuations of the resting membrane potentials distinguish certain CNS neurons from those in the spinal cord (compare with Fig. 1–4*D*). These fluctuations have been recorded from some, but not all, cells in the cerebral cortex, hippocampus and reticular formation, and unlike the EPSP's and IPSP's of the cord, these postsynaptic potentials do not seem to be related to the excitability of the cell as reflected in a spike discharge. In immature cortex not yet covered by contacts with other cells, multiple locations of rapid inflection points (arrows) indicate multiple trigger zones; note that spikes can arise from hyperpolarizations (from Purpura et al., 1965). In mature cortex, spikes arising from different critical firing levels (horizontal dashed lines) indicate "remote" excitation (from Elul and Adey, 1966). *D,* Illustration of results implying dendritic trigger zones (DTZ in *B*). A fast prepotential (FPP) recorded intracellularly is produced by a normal orthodromic stimulus arriving through the synaptic contacts. The FPP is not produced by an antidromic impulse initiated in the cell axon and sent backward toward the soma. The rapid inflection (arrows) indicating a trigger zone in the initial segment (IS in *B*) is produced by both types of stimuli. Spencer and Kandel (1961) interpreted this finding to mean that the trigger zone generating the FPP was not located in or near the initial segment and soma, since it would have been invaded by the antidromic action potential and would have appeared in the record. They argued that the FPP trigger zone was located out in the dendrites where the antidromic impulse could not invade it.

served. It was then interpreted that the FPP was produced by a trigger zone in the dendrites which was not invaded by the antidromic action potential. In immature cortical cells, which are not yet completely developed and covered by contacts with other cells, Purpura et al. (1965) observed inflection points at various positions on the action potentials recorded intracellularly from the same cell (see Fig. 1-5 *C*, arrows). Similar to the interpretation of the inflection point between the IS and SD discharge of VR cells (see Fig. 1-4 *C*), these various inflection points could be interpreted as signs of multiple patches of lower threshold trigger zones located at various sites within the cell. In the adult animal these multiple inflection points are not observed; the multiple threshold trigger zones have either become more remote or have disappeared. The initiation of propagated action potentials in the dendrites of cerebellar Purkinje cells has been demonstrated by measuring the extracellular potentials in the dendritic fields (Llinas et al., 1968). These results indicate that action potentials are generated in the dendrites and propagate toward the soma. It has also been shown that these traveling potentials can be inhibited or stopped before they reach the soma and are transferred to the axon.

Many cells in the central nervous system have complex postsynaptic potentials that may influence various trigger zones within the cell, each of which is capable of producing an action potential and transmitting information on to another cell. In many cases the postsynaptic potentials that are recorded by an intracellular microelectrode do not show the remote potentials that are in control of the various trigger zones. A similar instance has been noted in spinal cord VR cells, in which remote inhibition changed the cell's excitability but did not produce a recorded IPSP. The general principles of neuronal function which Eccles and his associates established in spinal cord motoneurons, namely, that EPSP's and IPSP's control the excitability of low threshold trigger zones, seem to be valid for the neurons of the central nervous system. However, the recorded activity inside the central neuron is much more complicated than that in the typical spinal cord motoneuron because the possibility for remote inhibition, remote excitation and remote trigger zones is greater, a condition no doubt related to the more complex structure and function of the elements of the central nervous system.

ELECTROGENESIS OF EEG POTENTIALS

Soon after it was discovered that weak electrical potentials could be recorded from the scalp of man (Berger, 1929), it was speculated that the waves which constituted the electroencephalogram (EEG) were the sum of the action potentials of individual neural elements (Adrian and Matthews, 1934, 1936). This hypothesis about the origin, or electrogenesis, of the EEG was based upon what was known at that time about action potentials in peripheral nerves, and it was accepted for many years. In 1953, however, Li and Jasper found that action potentials or spikes recorded from a small extracellular electrode could be abolished by mild anoxia without eliminating the EEG. It was then speculated that the EEG was the result of action potentials as well as other electrical potentials such as slow dendritic waves (Bishop, 1956; Bremer, 1958) and postsynaptic potentials (Purpura, 1959). Recent evidence indicates that the EEG may be the sum of the extracellular potentials produced by numerous, small, slow wave generators located on the membrane (Elul, 1968).

By inserting two microelectrodes into the brain and bringing their tips closer and closer together, Elul (1962) found that no correlation existed between the EEG activity recorded by either electrode when the tips were as close together as 30 μ; the electrodes would record the same activity if they were in contact with the same generator. This small distance would indicate that the generators of the EEG were smaller than a single cell. Elul (1968) found that the endogenous fluctuations of the membrane potential within a single cell contained the same frequency components (Fourier sine-wave components) as did the extracellular EEG, even when the EEG changed frequencies as it often does between states of drowsiness and wakefulness. It was suggested that these generators of the intracellular fluctuations may be the generators of the EEG. If the current of these generators flows into the cell through the site of generation, then out through the high resistance membrane, and then back through the extracellular space to the site of generation to complete the circuit, the potential fields created in the extracellular space will be determined by (1) the potential amplitude of the generator, and (2) the relative resistances of the cell membrane and the extracellular space. The amplitude of the intracellular membrane fluctuations is approximately 10 mV, and the cell membrane resistance per unit area is approximately 100 times greater than that of the extracellular space (Kuffler and Potter, 1964); hence, the computation for the extracellular field potentials is approximately 0.1 mV (see Fig. 1–8 C), which is the same order of magnitude as the actual recorded EEG.

If the EEG electrode is recording the potential field of more than one generator, which surely it must be, then the phase relation between these generators would be important in determining the amplitude of the recorded EEG. For example, if the generators are in phase with one another, i.e., if each one is producing extracellular currents flowing in the same direction at the same time in the vicinity of the extracellular recording electrode, then the EEG electrode will record a potential that is the sum of amplitudes produced by each generator. However, if the generators are out of phase with one another, and the potentials of each of the individual generators are normally distributed in time, then the EEG electrode will record an average potential that is much smaller and approaches the amplitude of one generator multiplied by the square root of the number of generators. The work of Elul (for review, see Elul, 1968) has indicated that indeed the potentials of these small generators are usually out of phase with one another and are normally distributed in time. During the special condition of slow wave synchrony of the EEG, which is usually associated with states of inattention or drowsiness (Fig. 1-6), a correlation can be found between the intracellular membrane fluctuations of a single cell and the recorded extracellular EEG. The EEG is found to increase in amplitude by about 10 times, but the amplitude of the potentials recorded inside a single cell increases only about twice (Elul, 1968). This increase in the membrane potential is not enough to account for the increase in the EEG, which suggests that the neuronal generators have, in addition, become somewhat synchronized and some of them are producing extracellular potentials in phase with each other.

During slow wave synchrony, and only during this condition, a

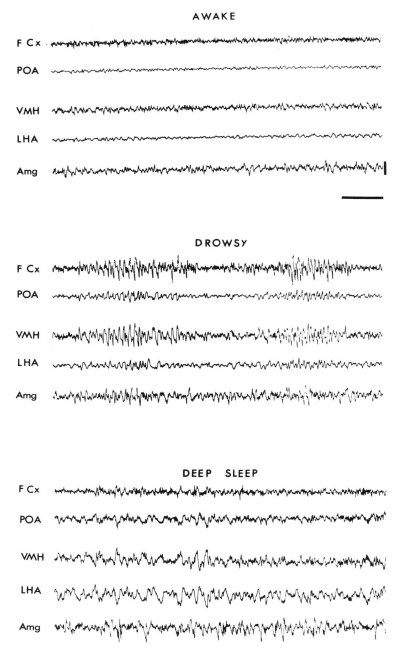

Figure 1–6.   EEG during wakefulness, drowsiness and sleep in the rat. EEG recorded from the cortex and various subcortical regions of the brain during different states of arousal from wakefulness to deep sleep. All recordings are from stainless steel bipolar electrodes. F Cx, frontal cortex; POA, preoptic area; VMH, ventromedial nucleus of the hypothalamus; LHA, lateral hypothalamic area; Amg, amygdala. Awake: Note widespread EEG desynchronization recorded during normal awake state. Drowsy: A few seconds later the animal became quiet and relaxed, and at most electrode sites the EEG showed spindle-shaped, 8 to 12 c/sec synchronous bursts of activity of varying amplitudes. Deep sleep: A few minutes later the animal was deeply asleep, with its head resting on the floor, and the EEG showed widespread, high voltage, 3 c/sec slow waves. Calibrations: 100 μV, 1 sec.

A

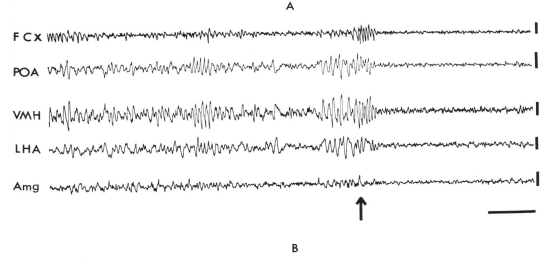

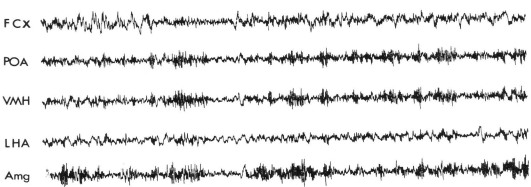

Figure 1–7 EEG during alerting and investigatory behavior. EEG recorded from same animal as in Figure 1–6. *A,* Desynchronization of the EEG of a drowsy animal produced by a loud noise (arrow). *B,* Alerting behavior, followed a few seconds later by sniffing and investigatory behavior. Note appearance of 40 c/sec activity in POA, VMH, and Amg. Calibrations: 100 μV, 1 sec.

correlation can be found between the spike discharge of a single unit and the EEG (Andersen and Eccles, 1962; Frost and Gol, 1966; Fox and Norman, 1968). The significance of this correlation between single unit action potentials and the EEG is not clear at present, but it does appear that such correlations occur only during EEG synchronization and do *not* occur in tissue that shows the low voltage desynchronized EEG pattern (Figs. 1–6 and 1–7).

RECORDING BIOELECTRIC POTENTIALS

The role of the extracellular space in the electrogenesis of the EEG is of paramount importance, as the return currents from active patches of membrane must flow through this channel. A model cell with an active generator site at the shaded negative pole is shown in Figure 1-8 *A*. The negative pole would normally draw current from all parts of

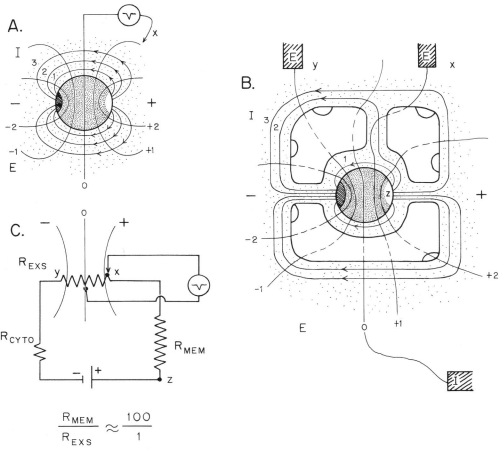

$$\frac{R_{MEM}}{R_{EXS}} \approx \frac{100}{1}$$

Figure 1–8 Model neuron. *A,* Diagrammatic representation of a single, active, depolarized patch of membrane (darkly shaded) on a model neuron, which is drawing current from many other parts of the cell membrane, but for purposes of illustration is represented as drawing current from a smaller equivalent patch (unshaded). Current flows from the positive to the negative pole through many symmetrically graded current pathways when the neural dipole is embedded in a homogeneously conductive medium. Most of the current flows through the shorter pathway (e.g., I_1) because it is the path of least resistance. Current also flows through the other, higher resistance pathways (e.g., I_2 and I_3) but in progressively smaller and smaller amounts. Associated with the currents flowing in the extracellular medium are electrical potentials. The values of these potentials will be equal at many points in the extracellular field, and they will fall on equipotential lines which are at right angles to the current paths. The equipotential lines which are of significant value will form closed loops within the vicinity of the poles, while the very small and insignificant potentials will fall on loops that are very large and extend far away from the dipole generator (the zero-potential line extends to infinity). *B,* Model neuron generator embedded in a homogeneously conductive medium which contains nonconductive bodies such as other neural and non-neural cells. Since the resistance of the nonconductive cell membranes is approximately 100 times greater than that of the extracellular conducting medium, most of the current flow between the poles of the model generator will course through the extracellular space. This will naturally change the shape of the equipotential lines which are orthogonal to the current paths. Since no appreciable amount of current flows through the nonconductive cells, no appreciable potentials will exist inside them (note that the dashed lines show the equipotential lines going around, not through, the cells). The recording electrode (E) placed at position X will record a positive potential with respect to the indifferent electrode (I), which is placed at a distance from the generator on the zero-potential line. If the electrode is moved to position Y, it will record a negative potential. *C,* Equivalent circuit for the model neuron generator. Current from the generator (represented by the battery) must flow through three different resistances: that of the membrane, R_{MEM}; the extracellular space, R_{EXS}; and the cytoplasm inside the cell, R_{CYTO}. The latter is insignificant in value with respect to the other two. R_{MEM} is 100 times greater than R_{EXS}, which means that the voltage recorded intracellularly (at point Z) will be 100 times greater than that recorded extracellularly (at point X or Y).

the cell membrane, but for purposes of illustration, this source can be represented by an equivalent pole as illustrated by the unshaded patch at the positive pole in the figure. As current flows from the positive to the negative pole through the homogeneously conducting extracellular space, a uniform electrical field is created along equipotential lines that are at right angles to the uniform current paths. Only the zero equipotential line, which extends to infinity, reaches an appreciable distance away from this model neuronal dipole; the other equipotential lines form closed loops within the vicinity of the cell. When this model cell is embedded in a group of other cells, as illustrated in Figure 1-8 B, then the current lines will be forced by the high resistance of the membranes of the other cells to travel in nonuniform channels. This situation will naturally change the shape of the equipotential lines, as illustrated.

When recording from this embedded model cell, the indifferent electrode (I) is assumed to be on the zero equipotential line, as this is the only one that extends away from the generator site. When the electrode (E) is moved from position X to Y along the dipole axis, the recorded potentials will change in polarity, as shown (Fig. 1-8 B). The equivalent electrical circuit for the neural dipole is shown in Figure 1-8 C. Note that placement of the recording electrode at point Z, inside the cell, will result in the recording of a potential that is 100 times greater than that recorded at either point X or point Y. The membrane resistance per unit area is 100 times greater than that of the extracellular space; therefore, the voltage drop across the former will be 100 times greater than that of the latter.

The effect of making an artificial current channel from an active generator site to the recording electrode is illustrated in Figure 1-9 A. An electrode is inserted step by step downward through the brain toward the optic tract (TO). At site 1, which is 2 mm above the TO, no potential is evoked by a flash of light directed into the animal's eye. However, if the electrode is driven into the TO and then returned to site 1', leaving a punctured channel connecting TO to site 1', then an evoked potential is recorded. (Care was taken to ensure that the electrode did not drag the TO back to site 1'.) A similar result is seen in Figure 1–9 C in which a punctured channel permits the recording of an active site distant to the recording electrode tip. The electrode was first driven down through the visual cortex and then moved up to a position in the middle of the cortex. Comparison of the records from site 3 and 3' indicates that a potential is channeled upward from the bottom of the puncture. Note that the potentials from below, not above, are channeled to the electrode; e.g., the negative polarity of the fourth peak recorded from site 7, not the positive polarity recorded from site 1, is channeled to site 3'. Apparently as long as the electrode remains in the punctured hole, the artificial current channel remains closed, permitting current conduction only from the unfilled puncture below. This conclusion is also supported by the record in Figure 1-9 B, which shows no conduction from site 4 to site 5 because the electrode is in the punctured channel and fills the space.

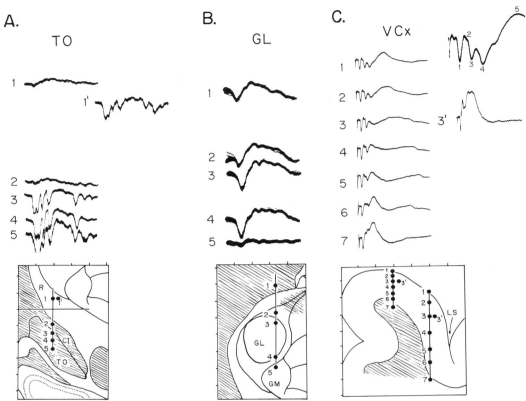

Figure 1–9 Evoked potentials. *A,* Potentials evoked in the optic tract (TO) after a brief light flash directed into the eye of an anesthetized cat. Records were taken at each one half millimeter step as the electrode was driven down toward the optic tract, as shown in the drawing. No evoked potentials were recorded until the electrode touched the TO at site 3. However, during withdrawal of the electrode after a penetration to site 5, an evoked potential was recorded 2 mm above the TO at site 1'. See text for discussion. *B,* Potentials evoked in the lateral geniculate body (GL) after a brief light flash directed into the eye. Recording sites are indicated in the drawing. Site 1 is in the radiation of optic fibers exiting the GL and traveling toward the visual cortex. Note that no response is recorded at site 5. Compare this to site 1' in A, in which the punctured channel between the generator of the visual evoked response and the recording site was not filled by the recording electrode. *C,* Potentials evoked in the lateral gyrus of the visual cortex (VCx) after a brief electrical stimulus to the optic tract. The four positive downward peaks followed by a fifth negative upward one are labeled in the extreme upper right trace. The stimulus artifact is shown in all traces as a brief downward deflection at the extreme left of each trace. Records essentially the same were obtained in a penetration perpendicular to the cortex and in one traveling obliquely through it, as indicated by the two tracks in the drawing. This implies a uniformity in wave form of the evoked potentials in horizontal lamina parallel to the surface of the cortex. Site 3' is the point of recording during withdrawal of the electrode after it has penetrated to site 7. Note that peaks 4 and 5 reverse in polarity between sites 3 and 4. Peak 3 reverses in polarity between sites 6 and 7. Peaks 1 and 2 do not reverse in polarity. See text for discussion. Abbreviations: CI, capsula interna; GL, geniculatus lateralis; GM, geniculatus medialis; LS, lateral sulcus; R, nucleus reticularis thalami; TO, tractus opticus; VCx, visual cortex.

It appears that an electrode as close as 0.5 to 0.25 mm from either the optic tract (Fig. 1-9 A, site 2) or lateral geniculate nucleus (Fig. 1-9 B, site 5) will not show an evoked response to light flash, presumably because there is no extracellular current channel connecting the active neural tissue to the recording electrode. If an evoked potential is recorded, then current from the generator must flow to the electrode either through an artificial channel created by the electrode or through a natural channel created in the extracellular space.

The direction and magnitude of the current flow at the tip of the electrode are important in determining the amplitude and wave form of the recorded potential. The change in polarity of the recorded potential which occurs when the electrode is moved from point X to Y along the axis of the generator dipole is shown in Figure 1–8 B. In the X position the current is flowing away from the recording electrode with respect to the indifferent electrode, creating a positive potential; in the Y position the current is flowing toward the electrode with respect to the indifferent electrode, creating a negative potential. If the distance between the poles of the generator were much greater, then moving the electrode the same distance from X to Y would not result in a change in polarity because the electrode would still be in the positive part of the field; however, a reduction in the amplitude might occur if the X-to-Y distance were great enough relative to the dipole length.

The effect of moving the recording electrode along the parallel axes of several generators whose dipoles are of different sizes is shown in Figure 1-9 C. The complex recorded potential, consisting of four positive peaks followed by a fifth negative one of longer duration, is produced in the visual cortex by electrical stimulation of the optic tract. Note in the numbered trace that the second positive peak is very small and appears as only a slight downward notch between peaks 1 and 3. As the electrode is lowered, there are reversals in polarity of some of the peaks. At site 3, peaks 4 and 5 have become very small and reverse in polarity simultaneously when the electrode is moved 0.5 mm to site 4. At site 6, peak 3 has become very small, and at site 7 it is reversed in polarity. Note that peaks 1 and 2 do not change in polarity even when the electrode has gone entirely through the cortex into the white matter.

The zero potential line, or midpoint, between the poles of the generators for peaks 4 and 5 is located between sites 3 and 4. This indicates that these short dipole generators are located within the cortex. For peak 3, the midpoint is at site 6, indicating a longer generator dipole. The poles of the generators for peaks 1 and 2 must be very far apart, because these potentials have not changed during the movement of the electrode through the cortex. It has been suggested by several investigators that the electrogenesis of the first peak results from activity in the presynaptic axons of the geniculocortical cells, whereas the electrogenesis of peak 4 is thought to come from activity in the soma and apical dendrites of the cortical cells. (See Brindley, 1960, p. 116, for review; also see Landau et al., 1964.) Peak 1 is produced by a generator extending from the thalamus to the cortex, whereas peak 4 is produced by a generator located wholly within the cortex.

REFERENCES

Adey, W. R., Kado, R. T., McIlwain, J. T., and Walter, D. O. (1966). Regional cerebral impedance changes in alerting, orienting and discriminative responses; the role of neuronal elements in these phenomena. *Exp. Neurol.* 15:490.

Adrian, E. D., and Matthews, B. H. (1934). The interpretation of potential waves in the cortex. *J. Physiol. (London)* 81:440.

Adrian, E. D., and Matthews, B. H. (1936). The spread of activity in the cerebral cortex. *J. Physiol. (London)* 88:127.

Andersen, P., and Eccles, J. C. (1962). Inhibitory phasing of neuronal discharge. *Nature (London)* 196:645.

Berger, H. (1929). Über das Elektrenkephalogramm des Menschen. *Arch. Psychiat. Nervenkr.* 87:527.

Bishop, G. H. (1956). Natural history of the nerve impulse. *Physiol. Rev.* 36:376.

Bremer, F. (1958). Cerebral and cerebellar potentials. *Physiol. Rev.* 38:357.

Brindley, G. S. (1960). *Physiology of the Retina and the Visual Pathway.* London, Edward Arnold.

Cajal, S. Ramon y (1911). *Histologie du Système Nerveux de l'Homme et des Vertébrés.* Paris, Maloine.

Cajal, S. Ramon y (1928). *Degeneration and Regeneration of the Nervous System.* Translated and edited by R. M. May. New York, Hafner Publishing Co., 1959.

Calvet, J., Calvet, M. C., and Scherrer, J. (1964). Étude stratigraphique corticale de l'activité EEG spontanée. *Electroenceph. Clin. Neurophysiol.* 17:109.

Creutzfeldt, O. D., Fuster, J. M., Lux, H. D., and Nacimiento, A. C. (1964). Experimenteller Nachweis von Beziehungen zwischen EEG-Wellen und Aktivitat corticaler Nervenzeilen. *Naturwissenschaften* 51:89.

Dale, H. H. (1935). Pharmacology and nerve-endings. *Proc. Roy. Soc. Med.* 28:319.

Dale, H. H. (1937). Transmission of nervous effects of acetylcholine. *Harvey Lect.* 32:229.

Eccles, J. C. (1964). *Physiology of Synapses,* Berlin, Springer-Verlag.

Elul, R. (1962). Dipoles of spontaneous activity in the cerebral cortex. *Exp. Neurol.* 6:285.

Elul, R. (1964). Specific site of generation of brain waves. *Physiologist* 7:127.

Elul, R. (1968). Brain waves: Intracellular recording and statistical analysis help clarify their physiological significance. *Data Acquisition and Processing in Biology and Medicine* 5:93.

Elul, R., and Adey, W. R. (1966). Instability of firing threshold and "remote" activation in cortical neurones. *Nature (London)* 212:1424.

Fox, S. S., and Norman, R. J. (1968). Functional congruence: An index of neural homogeneity and a new measure of brain activity. *Science* 159:1257.

Frank, K. (1959). Basic mechanism of synaptic transmission in the central nervous system. *IRE Trans. Med. Electronics* ME 6:85.

Frank, K., and Fuortes, M. G. F. (1957). Presynaptic and postsynaptic inhibition of monosynaptic reflexes. *Fed. Proc.* 16:39.

Frost, J. D., Jr., and Gol, A. (1966). Computer determination of relationships between EEG activity and single unit discharges is isolated cerebral cortex. *Exp. Neurol.* 14:506.

Fujita, J., and Sato, T. (1964). Intracellular records from hippocampal pyramidal cells during theta rhythm activity. *J. Neurophysiol.* 27:1011.

Furshpan, E. J., and Potter, D. D. (1968). Low-resistance junctions between cells in embryos and tissue culture. *Current Topics in Developmental Biology* 3:95.

Galambos, R., and Morgan, C. T. (1960). The neural basis of learning. *In* Fields, J. C. (ed.): *Handbook of Physiology-Neurophysiology, III.* Washington, D.C., American Physiological Society.

Gasser, H. S. (1939). Axons as samples of nervous tissue. *J. Neurophysiol.* 2:361.

Granit, R., Kellerth, J. O., and Williams, T. D. (1964). "Adjacent" and "remote" postsynaptic inhibition in motoneurons stimulated by muscle stretch. *J. Physiol. (London)* 174:453.

Hodgkin, A. L. (1964). *The Conduction of the Nervous Impulse.* Springfield, Ill., Charles C Thomas.

Hodgkin, A. L., and Huxley, A. F. (1952). A quantitative description of membrane current and its application to conduction and excitation in nerve. *J. Physiol. (London)* 17:500.

Hydén, H. (1959). Biochemical changes in glial cells and nerve cells at varying activity. *In* Hoffmann-Ostenhoff, O. (ed.): *Biochemistry of the Central Nervous System. Vol. III.* Proceedings of the Fourth International Congress of Biochemistry. London, Pergamon Press, p. 64.

Jasper, H. J., and Stefanis, C. (1965). Intracellular oscillatory rhythms in pyramidal tract neurons in the cat. *Electroenceph. Clin. Neurophysiol.* 18:541.

John, E. R. (1967). *Mechanisms of Memory.* New York, Academic Press, Inc.

Kandel, E. R., Frazier, W. T., and Coggeshell, R. (1966). Opposite synaptic actions mediated by different branches of an identifiable interneuron in Aplysia. *Fed. Proc.* 25:456.

Krech, D., Rosenzweig, M. R., and Bennett, E. L. (1963). Effects of complex environment and blindness on rat brain. *Arch. Neurol. (Chicago)* 8:403.

Kuffler, S. W., and Potter, D. D. (1964). Glia in the leech central nervous system: Physiological properties and neuro-glia relationship. *J. Neurophysiol.* 27:290.

Landau, W. M., Bishop, G. H., and Clare, M. H. (1964). Analysis of the form and distribution of evoked cortical potentials under the influence of polarizing currents. *J. Neurophysiol.* 27:788.

Lashley, K. S. (1930). Basic neural mechanisms in behavior. *Psychol. Rev.* 37:1.

Lashley, K. S. (1950). In search of the engram. *Sympos. Soc. Exp. Biol.* 4:454.

Li, C.-L., and Jasper, H. H. (1953). Microelectrode studies of the electrical activity of the cerebral cortex in the cat. *J. Physiol. (London)* 121:117.

Ling, G. N. (1964). The association-induction hypothesis. *Texas Rep. Biol. Med.* 22:244.

Ling, G. N. (1965). Physiology and anatomy of the cell membrane: The physical state of water in the living cell. *Fed. Proc.* 24:103.

Ling, G. N. (1969). A new model for the living cell: A summary of the theory and recent experimental evidence in its support. *Int. Rev. Cytol.* 26:1.

Llinas, R., Nicholson, C., Freeman, J. A., and Hillman, D. E. (1968). Dendritic spikes and their inhibition in alligator Purkinje cells. *Science* 160:1132.

McGaugh, J. L. (1965). Facilitation and impairment of memory storage processes. *In* Kimble, D. P. (ed.): *The Anatomy of Memory.* Palo Alto, Calif., Science and Behavior Books, Inc., p. 240.

Müller, J. (1838). *Handbuch der Physiologie, II,* bk. V. Out of print.

Purpura, D. P. (1959). Nature of electrocortical potentials and synaptic organizations in cerebral and cerebellar cortex. *Int. Rev. Neurobiol.* 1:47.

Purpura, D. P., Shofer, R. J., and Scarff, T. (1965). Properties of synaptic activities and spike potentials of neurons in immature neocortex. *J. Neurophysiol.* 28:925.

Reisen, A. H. (1961). Studying perceptual development using the technique of sensory deprivation. *J. Nerv. Ment. Dis.* 132:21.

Robinson, C. E. (1966). A chemical model of long term memory and recall. *In* Walaas, O. (ed.): *Molecular Basis of Some Aspects of Mental Activity. Vol. I.* New York, Academic Press, Inc., pp. 29–35.

Salmoiraghi, G. C., and Stefanis, C. N. (1967). A critique of iontophoretic studies of central nervous system neurons. *Int. Rev. Neurobiol.* 10:1.

Scheibel, M. E., and Scheibel, A. B. (1966). Patterns of organization in specific and nonspecific thalamic fields. *In* Purpura, D. P., and Yahr, M. (eds.): *The Thalamus.* New York, Columbia University Press.

Scheibel, M. E., and Scheibel, A. B. (1967). Structural organization of nonspecific thalamic nuclei and their projection toward cortex. *Brain Res.* 6:60.

Segundo, J. P., Takenaka, T., and Encabo, H. (1967). Electrophysiology of bulbar reticular neurons. *J. Neurophysiol.* 30:1194.

Shannon, C. E. (1949). The mathematical theory of communication. *In* Shannon, C. E., and Weaver, W.: *The Mathematical Theory of Communication.* Urbana, University of Illinois Press, pp. 3-93.

Sherrington, C. S. (1906). *The Integrative Action of the Nervous System.* London, Cambridge University Press.

Spencer, W. A., and Kandel, E. R. (1961). Electrophysiology of hippocampal neurons. IV. Fast prepotentials. *J. Neurophysiol.* 24:272.

Sperry, R. W. (1944). Optic nerve regeneration with return of vision in anurans. *J. Neurophysiol.* 7:57.

Sperry, R. W. (1965). Embryogenesis of behavioral nerve nets. *In* DeHaan, R. L., and Ursprung, H. (eds.): *Organogenesis.* New York, Holt, Rinehart, and Winston.

Szentagothai, J., and Rajkovits, K. (1955). Die Ruckwirkung der spezifischen Funktion auf der Struktur der Nervenelemente. *Acta Morph. Acad. Sci. Hung.* 5:253.

Terzuolo, C. A., and Bullock, T. H. (1956). Measurement of imposed voltage gradient adequate to modulate neuronal firing. *Proc. Nat. Acad. Sci.* 42:687.

Ungar, G. (1970). Molecular mechanisms in information processing. *Int. Rev. Neurobiol.* 13:223.

Wang, H. H., Tarby, T. J., Kado, R. T., and Adey, W. R. (1966). Periventricular cerebral impedance after intraventricular calcium injection. *Science* 154:1183.

Wiener, N. (1948). *Cybernetics.* New York, John Wiley & Sons, Inc.

Wiesel, T. N., and Hubel, P. H. (1963). Single-cell responses in striate cortex of kittens deprived of vision in one eye. *J. Neurophysiol.* 26:1003.

CHAPTER 2

FUNCTIONAL NEUROANATOMY
AND DISSECTION GUIDE
FOR THE COW AND SHEEP BRAINS

Along the phylogenetic scale of the vertebrate animals three distinct divisions of the cerebrum of the brain can be observed: (1) *mesencephalon*, composed of the tectum (superior and inferior colliculi), underlying tegmentum (reticular formation) and basis pedunculi (ascending sensory and descending motor fibers), and present in all vertebrates; (2) *diencephalon*, composed of the thalamus, subthalamus and hypothalamus, and present in all vertebrates but more developed in the higher ones; and (3) *telencephalon*, composed of the cortex and basal ganglia (caudate, globus pallidus, amygdala, claustrum, hippocampus, basal forebrain area), and present in completely developed form only in the higher vertebrates. The cerebrum plus the cerebellum, pons and medulla constitute the entire central nervous system. The cerebrum is the most highly developed part of the brain and is thought to be composed of several types of brain systems which underlie the higher nervous functions of conscious perception and motor reaction. The other parts of the brain are related largely to the unconscious homeostatic and reflex functions of the organism, such as control of vital body functions, smoothing out muscular movements and maintaining postural balance.

A brain system is defined in terms of its correlated anatomical, physiological, and behavioral characteristics. The anatomical boundaries of a system are delineated by the related physiological and behavioral correlates, as can be seen, for example, in the visual system, in which the anatomical structures are those that manifest electrical evoked potentials and produce behavioral blindness when lesioned or destroyed. Each of these three types of observations can be made without reference to the others, but when they are brought together, naming the parts, mechanism and actions of the structure, then the concept of "system"

emerges. No brain system is self-contained, and each operates different-
ly in the intact brain than it does after surgical isolation. However,
there are certain conceptually distinct features of the functions of various regions of the brain that enable basic anatomical divisions to be
outlined.

TABLE 2–1 BRAIN SYSTEMS

Common Name	Cortical Component	Inferred Higher Nervous Function
Specific Systems		
Cutaneous sensory system	Sensorimotor cortex	Perception of skin senses (cold, heat, touch, pressure, pain)
Proprioceptive system		Perception of limb position and movement
Motor system		Voluntary movement of limbs
Visual system	Visual cortex	Perception of visual forms and movements
Auditory system	Auditory cortex	Perception of auditory sounds
Olfactory system	Olfactory cortex	Perception of odors
Association Systems		
Parietal lobe association system	Parietal cortex	Associative perception of specific senses involving learning and memory
Temporal lobe association system	Temporal cortex	Associative perception same as above and also perception of emotional and specific sensory interactions
Frontal lobe association system	Frontal and sensorimotor cortex	Associative perception and sensory-motor interactions
Nonspecific Systems		
Reticular activating system	Unknown; however, some projections to frontal cortex	Reflexive control of wakefulness, arousal, and specific sensory systems (attention)
Nonspecific thalamo-cortical system	Frontal cortex	Voluntary control of specific and associative sensory systems (selective perception) and learned behavior patterns (selection among strong response tendencies)
Emotion Systems		
Limbic system (rhinencephalon; visceral brain; tertiary olfactory system)	Cingulate and hippocampal cortex	Sensor of emotional awarenesses (pleasure, fear, rage, sex, hunger, etc.) and motivational processes
Hypothalamic effector mechanisms	Unknown, but most higher systems project into hypothalamic regions	Effector of emotional motor reactions, and effector and sensor for autonomic nervous system

Table 2-1 shows the common names, cortical representations and higher nervous functions of the brain systems. The cortex is the most advanced component of each of the systems and is a distinguishing feature of most of the systems of interest. Figures 2-18 to 2-23 show the anatomical boundaries of the various systems.

SPECIFIC SYSTEMS

In 1838 Johannes Müller published his theory of specific nerve energies, in which he stated that the sensorium knows directly only the state of the nerves; these nerves have specific irritabilities and, thus, provide the necessary information for the sensorium to assess the external world of stimulation. In other words, we perceive in a particular sense modality because that particular nervous system is being stimulated. For example, stimulation of the visual system at any point, such as the retina, optic nerve, lateral geniculate or visual cortex, will always produce a visual sensation because of the specificity of the nerves being stimulated.

Each of the specific sensory systems has a peripheral pathway leading from its sensory receptors to a specific part of the thalamus. The thalamic nucleus then relays the information to a cortical region, called a primary cortical projection zone, that is located in a separate part of the cortex for each sense modality. In summary, each specific system has connections throughout its length which are separate from the others and specific to its particular sense modality.

The olfactory system does not relay in the thalamus as do all of the other sense modalities. This particular specific system is closely related to a part of the brain, called the limbic system, present in all rudimentary to advanced vertebrate forms; it will be discussed later.

The voluntary motor system is not one of the specific sensory systems but is included here in this discussion because of its close anatomical and physiological relationship to the cutaneous or skin-sense system at the level of the cortex. Together they are commonly called the sensorimotor system. The thalamic relay nucleus of the motor component of the system (ventralis lateralis) is in the same ventral nuclear group of the thalamus as that of the cutaneous sensory one (ventralis posterolateralis and ventralis posteromedialis), confirming their close anatomical relationship. The motor relay nucleus receives a major input from the cerebellum via the brachium conjunctivum and relays this input to the motor cortex. The motor cortex also receives major input from the cutaneous sensory cortex, then sends descending axons back through the cerebral peduncles and pyramidal tracts to end upon cells of the propriospinal system in the cord in order to produce complex limb movements. In primates many of these descending motor fibers also end upon the ventral root motor neurons which innervate the limb muscles directly and seem to provide for fine muscle control. The cutaneous sensory relay nucleus of the thalamus receives its input from the medial lateral lemnisci and relays the somatic sensory information from these fibers to the cutaneous sensory cortex. The sensorimotor system underlies *integrated movements* produced by contractions of several muscles and does not control the contraction of individual muscles.

When a person directs his sensory receptors toward an object, he can distinguish between the simple sensations (feel, smell, visual form) and the more complex ideas (booklike, chairlike) related to the perception of that object. One of the early philosophies regarding conscious awareness maintains that these simple sensations and complex ideas are of the same perceptual kind, although varying in intensity and vividness, and most importantly, it states that both are learned by the processes of mental association which occur after conscious experience with the objects (Locke, 1690; Hume, 1739). In contrast to this early associationist view, Immanuel Kant (1781) held that there are certain *a priori* conditions for conscious awareness in addition to those *a posteriori* ones created from experience. For example, he points out that objects are not actually related in space and time, but the mental impressions from these objects must necessarily be encoded in a spatial and temporal relationship in order to be perceived. Johannes Müller was greatly influenced by Kant, and his physiologic concept of specific nerve energies is an example of an a priori condition necessary for conscious perception.

Support and clarification of Kant's view have been achieved in more recent times through a simple observation by Lettvin (1967), who found that the more rudimentary animal forms perceive objects in the external world in the same way as the most highly developed forms. He observed that among the leaf insects those that most resembled leaves survived better during forages of the insects' predators. These predators included a wide variety of animals with different perceptual mechanisms and brains which all see leaves and near-leaves in the same manner, thus permitting the natural selection of the most leaflike insects. This constancy in psychologic awareness indicates the presence of some a priori condition for perception which each of the different perceptual mechanisms follows.

Lettvin et al. (1959) and Maturana et al. (1960) first showed that there are neural elements of the frog's retina which respond selectively to certain visual stimuli such as sustained edges, moving convex edges, changing contrasts, dimming and darkness. Hubel and Wiesel (1959, 1962, 1963a, b, c, 1965) found single neural units in the cortex of the cat which detect selectively even more complex visual forms such as boundaries, line orientations, directions of movement, angles, sizes, positions and shapes. These studies have demonstrated that there are fixed neural machanisms which determine a priori what visual forms are processed into sensory information. Furthermore, the work of Hubel and Wiesel (1963b) strongly suggests that such visual mechanisms are innate and unlearned, because the above-mentioned detectors of visual form were found in young kittens without prior visual experience. They found, however, that some visual-form detectors were missing in adult animals deprived of form vision from birth, suggesting the degeneration of an innate perceptual mechanism in the brain which is not used.

Perceptual mechanisms for innate or unlearned sensations would be inconsistent with a purely associationist philosophy, whereas in a philosophy requiring a priori conditions for conscious awareness, certain perceptual mechanisms must be innate. Perhaps the historical distinction between innate-specific and learned-associative sensory processes made by the early philosophers continues to serve our neurosciences today, for there appear to be elements of both types of sensory mechanisms within the brain.

ASSOCIATION SYSTEMS

The thalamus can be divided into three parts, depending upon the nature of the input and output of its cells. The extrinsic thalamic nuclei receive most of their inputs from peripheral pathways and relay these inputs to the cortex. These nuclei comprise the ventral and lateral parts of the thalamus and are components of the specific systems. The medial thalamic nuclear group receives its input from many different sources, and projects upon many other thalamic nuclei, as well as the frontal cortex, via axonal branches (collaterals). These nuclei constitute the medial region of the thalamus and are part of the nonspecific systems. The third division of the thalamus contains the intrinsic nuclei. These nuclei are like the specific sensory relay nuclei which project to restricted regions of the cortex, but they do not receive their input from the peripheral sense organs. Rather, they receive their input from other thalamic nuclei, both specific and nonspecific, and are thought to integrate or associate several of the specific sensory systems with one another. These nuclei constitute the dorsolateral parts of the thalamus and project directly to one of several cortical association regions of the temporal, parietal and frontal lobes.

The historical distinction between the specific and the association systems was discussed previously. The specific systems seem to be innate sensory mechanisms which predetermine what stimulus forms from an object will be perceived, whereas the association systems seem to be involved with the more complex ideas in perception, relating sensory experiences in such a way that more than the simple sensations are perceived from the object. That is, learning, conditioning, memory and other higher order associative processes operate upon the predetermined stimulus input to produce an awareness later that was not possible before the experience.

The single neural elements of the cortical association regions are multisensory, although some of them do not respond to all sense modalities, only to singles, pairs or triplets of them. None of the individual neurons of the specific systems changes its response patterns with repeated experience of sensory stimuli, whereas many of those single cells investigated in the association systems do. Morrell (1967), for example, has shown that single cells in the association cortex surrounding the primary visual cortex will produce constant, but different, poststimulus firing patterns to either a light or sound presentation. If the light and sound are then temporally related by a repeated presentation of light followed by sound, then the light will begin to produce a different poststimulus-response pattern during the period before the sound is presented. This conditioned associative response will return to the original response pattern if the sound is no longer temporally related to the light.

Penfield (1958) has observed that electrical stimulation of epileptogenic tissue in the temporal association cortex in man produces higher order, "interpretative," sensory perceptions similar to hallucinations or vivid remembrances, whereas stimulation of the specific sensory cortices produces only specific sensory perceptions such as light flashes, spots, clicks and buzzes. The most commonly elicited perceptual awareness from direct stimulation of the association cortex is that of familiarity. No matter what the concurrent specific sensory perceptions, remembrances or hallucinations, they are accompanied, familiar or not, by an awareness of, "I know this, I have perceived all of this before."

Lesion or ablation studies of the association cortices in animals have revealed marked deficits in complex sensory discrimination tasks. The frontal, parietal and temporal assocation cortices are wedged between the primary sensory ones, and the effect of a particular ablation in the association cortex on sensory discrimination is usually more marked in the modality of the specific sensory cortex adjacent to the ablation, although other modalities are also affected (Chow and Hutt, 1953; Pribram, 1954). It has often been difficult to interpret whether these studies show association sensory deficits or merely specific sensory deficits, because if it is to be shown that an animal cannot discriminate between two associative aspects of a visual stimulus (e.g., relative position, roundedness or intensity), it must first be proved that the animal is not functionally blind and can still perceive the specific sensory aspects of these stimuli, some of which are quite complex.

Studies in man have helped somewhat in the interpretation of the function of the association cortex. Patients with damage in the temporal and parietal association cortices manifest certain interesting clinical symptoms (Nielsen, 1946): *apraxias*, deficits in movement patterns such as playing the piano or typing; *agnosias*, sensing but not knowing what an object is used for or what it does; and several types of *aphasias*, such as expressive aphasia, in which there are deficits in speech, except under certain types of emotional stress, but written and spoken words are understood and writing is unimpaired; receptive aphasia, in which spoken or written words cannot be understood, amnesic aphasia, in which there is inability to relate nouns to objects but everything about the actual object is known; and semantic aphasia, in which the individual cannot comprehend complex sentence structure. In none of these disorders is there paralysis, deafness, blindness or any kind of sensory disorder, but rather a loss of some higher order associative process relating the sensations and their interpretive meanings. Patients with damage in the frontal association cortex show cutaneous sensory-motor and visual-motor associative disorders (Milner, 1964; Teuber, 1964), but often such damage is combined with damage to the frontal granular cortex, which seems to involve a different kind of brain mechanism and produces a different type of functional disorder; this will be discussed in the next section.

NONSPECIFIC SYSTEMS

The nonspecific systems are related to perception and sensation but without the specificity to a single sense modality that characterizes the specific systems. The single neural units of these systems are multisensory, being activated by stimuli in all the specific sense modalities. They are called nonspecific because they are multisensory and are related to the specific senses in a manner exactly opposite to the meaning of "specific" given by Johannes Müller. For example, when we see, there must be a visual stimulus (specific), but we can be aroused from sleep by, or direct our attention to, any stimulus—a noise, a touch or a light flash (nonspecific).

The nonspecific systems are not related to the content of the specific sensations or the learned associative awarenesses but rather to the control of their admission into conscious perception, i.e., attention. One can fix his vision upon an object in the room and become aware of the details of that object while suppressing awareness of objects in the background, and then without moving his eyes he can become aware of the details of the objects in the background without further awareness of the fixated object. Or he can become aware of sounds and suppress what is seen, or concentrate on internal thoughts while eliminating awareness of external stimuli. However, if a novel sensation occurs or a meaningful stimulus is presented, he immediately becomes aware of it whether or not he wants to. According to William James (1890), attention (1) may be focused on either *external* or *internal* events (i.e., on objects or memories), (2) may be *innate* (i.e., immediately drawn to a certain quality of a stimulus) or *learned* (i.e., derived by association with other stimuli), and (3) may be *reflexive* or *voluntary*.

In each of these instances of attention that can be subjected to experimental observation, a common feature of brain activity has been observed: desynchronization of the EEG (e.g., see Figs. 1-6 and 1-7). A human subject who voluntarily directs his attention toward a visual stimulus will show desynchronization of the 8-12 c/sec alpha rhythm in the visual cortex but will not show desynchronization in this region if he attends to an auditory stimulus (Adrian, 1944; an example of external, learned, voluntary attention). A human subject instructed to perform mental arithmetic will show desynchronization of the alpha rhythm during the period of computation (Adrian, 1944; internal, learned, voluntary). A drowsy animal will show widespread electrocortical desynchronization to the presentation of a conditioned stimulus which was ineffective before conditioning (Morrell and Jasper, 1956; external, learned, reflexive). A drowsy animal will show desynchronization to the presentation of a novel stimulus (Rheinberger and Jasper, 1937; external, innate, reflexive). James' other classifications of attention have not been successfully observed experimentally because no way has yet been found to control an internal stimulus and to interpret what an innate, voluntary type of attention would be (James also had difficulty interpreting this latter category). Thus, there appears to be an underlying mechanism common to the several varieties or aspects of attention that is related to the mechanisms of electrocortical desynchronization.

The first unequivocal evidence for a brain mechanism responsible for electrocortical desynchronization was provided in a study by Moruzzi and Magoun (1949), who showed that electrical stimulation of the mesencephalic reticular formation causes widespread desynchronization of electrocortical activity. Stimulation of the reticular formation will awaken and arouse a sleeping or drowsy animal (Segundo et al., 1955) and will lower perceptual thresholds in the awake animal (Fuster, 1958). Lesion of this same region, in contrast, causes widespread electrocortical synchrony, manifested by slow waves and spindle bursts, and is accompanied by behavioral somnolence or sleep (Lindsley et al., 1949, 1950). The pathways by which the ascending reticular activating system achieves cortical desynchronization have not been clearly delineated, but because of lack of sufficient evidence for widespread reticulocortical projections and the abundance of evidence for reticulothalamic projections to the medial thalamic nuclear group (Brodal, 1957; Rossi and Zanchetti, 1957; Scheibel and Scheibel, 1958), the latter appear to constitute a major route. Therefore, electrocortical desynchronization appears to be achieved by reticular action upon a thalamocortical system, which in turn has widespread cortical influences.

The exact pathways mediating the synchronization and desynchronization effects of this medial thalamocortical system are not known at present, but it seems clear that the integrity of a midline thalamic system interconnected with the lateroventral and orbital cortex of the frontal lobe through the inferior thalamic peduncle is essential for the production of electrocortical synchrony. Lesions at any point in this latter nonspecific thalamocortical system will abolish induced or spontaneously occurring synchronization (Lindsley et al., 1949; Hanbery et al., 1954; Velasco and Lindsley, 1965; Skinner and Lindsley, 1967), whereas low frequency stimulation will produce such hypersynchronous activity (Dempsey and Morison, 1942a, b; Morison and Dempsey, 1942, 1943; Schlag et al., 1966). Low frequency stimulation of the nonspecific thalamocortical system appears to relax an animal and produce inattention, drowsiness and sleep (Hess, 1944; Hunter and Jasper, 1949; Akert et al., 1952; Akimoto et al., 1956; Buchwald et al., 1961; Sterman and Clemente, 1961; Buser et al., 1964). Thus it appears that the consequences of action of this nonspecific thalamocortical system on electrocortical synchronization and behavior are opposite to those of the reticular activating system.

During experimentally induced states of selective perception in both man and animals, the primary evoked response of the attended stimulus is enhanced, whereas that of an unattended stimulus in another sense modality is unaffected (Galambos et al., 1956; Hearst et al., 1960; Marsh et al., 1961; Galambos and Sheatz, 1962; Garcia-Austt, 1963; Garcia-Austt et al., 1964; Haider et al., 1964; Satterfield and Cheatum, 1964; Spong et al., 1965). Evoked potentials recorded in specific sensory cortex are enhanced in amplitude by either of two mechanisms: stimulation of the mesencephalic reticular formation (Bremer and Stoupel, 1959; Dumont and Dell, 1960; Narikashvili, 1963) or functional blockade of the nonspecific thalamocortical system (Skinner and Lindsley, 1967, 1971).

How the many types of attention suggested by William James are regulated by these two nonspecific systems of opposite neural action is not known; they converge in the medial thalamus, and both regulate electrocortical synchronization and excitability to sensory stimuli but by different mechanisms. The nonspecific thalamocortical system enhances sensory evoked potentials and reduces synchronous activity in the cortex by blocking or turning off its effects. In contrast, the reticular activating system produces the same results by increasing or turning on its effects. During partial blockade of the nonspecific thalamocortical system, enhancement of evoked potentials occurs in only one sense modality, whereas during complete blockade, widespread enhancement occurs in other specific and association cortices (Skinner and Lindsley, 1971). This finding indicates that the nonspecific thalamocortical system is capable of differential control of electrocortical excitability, a capacity one would expect for a neural mechanism underlying selective perception. On the other hand, stimulation of the mesencephalic reticular formation seems to produce a more generalized and widespread form of electrocortical desynchronization and sensory facilitation (Jasper, 1960; Narikashvili, 1963), an effect one would expect for a neural mechanism underlying reflexive perception.

The frontal granular cortex, a major part of the nonspecific thalamocortical system, constitutes one third of the cortical mass in the human brain, markedly distinguishing the brain of man from that of the lower animals. The mesencephalic reticular formation, on the other hand, is present in all rudimentary vertebrate forms. Pherhaps what we call human intelligence is explained in part by the particular types of attention presided over by the nonspecific thalamocortical system, whereas the reticular activating system underlies the more fundamental types of attention necessary for the more basic higher nervous functions.

Cortical ablation or lesion of the frontal granular cortex in animals does not result in an inability to perceive either specific or associative stimuli, but does produce an inability to perform discrimination among them because of the interfering influences of other operative perceptual mechanisms; that is, the animal cannot inhibit the effects of other intervening perceptual influences while it performs the discrimination task. Two particular types of tasks are especially disrupted by lesions of the frontal granular cortex: the delayed-response task and the discrimination-reversal task. In the former, the subject is required to withhold his discriminatory response for a brief time interval after he has received the relevant discrimination cues; the lesioned animal makes errors by perseveratively responding to certain sensory cues in the test situation (Malmo, 1942; Mishkin and Pribram, 1956; Brush et al., 1961; Mishkin et al., 1962). In the latter task the subject is required to discontinue his responding to a previously correct discriminative cue and respond instead to the previously incorrect one; the lesioned animal makes errors because it persists in responding to the previously correct cue (Brush et al., 1961; Mishkin et al., 1962). However, the frontal lobe-lesioned animals can perform both these tasks correctly if certain strong tendencies to respond to irrelevant aspects of the sensory cues are eliminated. If a tendency to perform a response to a preferred stimulus (usually the first correct one chosen in learning the task) is eliminated by nonreward of this response, then the animal can correctly perform the delayed-response task (Konorski and Lawicka, 1964). If a tendency to respond to the previously correct cue is eliminated by substituting a new stimulus in its place at the time of required response reversal, the animal quickly learns to respond to the previously incorrect stimulus (Mishkin, 1964). In summary, the lesioned animal can perform the complex discrimination tasks but is often unable by itself to choose among the various alternative cues that control his response behavior.

EMOTION SYSTEMS

Certain sensations and motor responses of which we are totally unaware occur within our bodies to regulate vital functions. The autonomic nervous system maintains homeostasis by responding to various changes in internal stimuli with an appropriate motor regulatory action. For example, changes in the content of the gases in the blood or changes in body temperature stimulate such homeostatic regulatory responses as an increase in respiration or vasoconstriction of the small, heat-radiating vessels in the skin. All these sensory and motor events occur automatically and without conscious awareness. Such homeostatic mechanisms are necessary for the survival of the organism and occur in an invariant environment where the sensory information is not complex and the motor responses are relatively simple.

Other homeostatic-like mechanisms are also necessary for the survival of the organism as well as for its species, in which the stimuli are quite variable and complex and the motor responses involve lengthy series of movements which differ in each instance. These are the survival behaviors of seeking food and mates, defending them, and escaping from predators. In order for these homeostatic-like, higher order behaviors to be carried out, the relevant stimuli from external and internal sources must be subject to conscious perception by the higher order specific, association and nonspecific mechanisms, so that adaptable responses can be made to them. For example, when an organism is aware of hunger, it must respond by an increase in food intake, but to accomplish this task it must first seek food and distinguish among the edible and inedible objects in its environment, behavior which requires the use of all the higher order, conscious faculties. The perceptions generated by these higher order faculties are subject to conscious awareness, as are the response-initiating sensations themselves, the emotional feelings. One can sense feelings of hunger, fear, sexual urges and rage which motivate or initiate the complex regulatory behaviors. MacLean (1949, 1958) has discussed the visceral origin of these feelings and has referred to their related brain mechanisms as the "visceral brain." It seems that these visceral-brain systems are well developed in rudimentary vertebrate forms and that they contain the motivational, discriminatory and motor mechanisms which the other higher order faculties are able to assist.

The mechanisms of conscious emotion are closely related to the homeostatic mechanisms that regulate the bodily needs. During the emotional state of fear, for instance, the sympathetic division of the autonomic nervous system produces vasoconstriction of the blood vessels in the skin, vasodilation of the blood vessels in the muscles, secretions of adrenaline from the adrenal glands, and other responses to mobilize the organism for immediate behavioral action relevant to the fear-evoking stimuli. Such an interaction is beneficial for the survival of the organism. However, this close connection can at times be harmful to the animal instead. Stressful emotional situations can cause a man to die suddenly from cardiac fibrillation produced by an otherwise minor irritation to the heart (Parkes, 1967; Wolf, 1969). An animal forced to make decisions to avoid an electric shock will develop gastric ulcers, whereas a randomly shocked control animal does not (Brady, 1958). The close interconnection between the two systems suggests that there is a primitive origin common for both. The nonbeneficial interactions may be through residual connections which remain as the conscious visceral regulatory mechanisms emerged from the unconscious autonomic ones.

William James (1890) and C. G. Lange (1885) thought that the conscious emotions were actually the sensations perceived from the sensors in the somatic and visceral tissues. According to their view, in a situation in which a man is running away from a large bear, he feels the emotion of fear because the reflexive running response causes excitation of the sensors in his muscles and viscera which then produces the feeling of fear. However, it has been shown that complete deafferentation of the sensory nerves of the autonomic and somatic nervous systems does not abolish emotional reactions in animals (Sherrington, 1900; Cannon, 1927, 1931) or emotional feelings in humans (Dana, 1921). Therefore, emotional sensations or behavioral reactions must arise from some source other than specific sensory perceptions from internal sensors.

Any type of regulatory mechanism has both sensors and effectors which can operate independently of one another. For example, if the thermostat is bypassed, the furnace in a heating system of a house can be turned on even though the house is hot. In like manner, the effector reactions of the brain can become operative by bypassing sensory receptors and directly stimulating the effectors of that regulatory mechanism. Direct stimulation of the hypothalamus can produce eating and drinking behavior in a satiated animal (Brugger, 1943; Andersson, 1951; Delgado and Anand, 1953; Hess, 1954; Larsson, 1954; Andersson et al., 1958; Akert, 1961; Smith, 1961), elicit spontaneous sexual activity (MacLean and Ploog, 1962; Vaughan and Fisher, 1962), and produce reactions of fear (Hess, 1936, 1949, 1954; Roberts, 1958a, b; Miller, 1961), rage (Hess, 1928, 1936, 1949, 1954; Ingram et al., 1936; Karplus, 1937; Ranson, 1939; Ranson and Magoun, 1939; Hinsey, 1940; Ingram, 1952; Hunsperger, 1956; Nakao and Maki, 1958; Roberts, 1958a, b; Akert, 1961) and pleasure (Olds, 1958, 1962).

Mounting evidence supports the view that the hypothalamus is a major effector organ that can produce overt behavioral reactions without the concomitant emotional sensations, just as stimulation of the sensorimotor cortex can produce movements of the limbs even though there is no preceding or accompanying sensory event. The rage reaction (Masserman, 1941) and eating behavior (Brugger, 1943; Hess, 1954; Akert, 1961) elicited by direct hypothalamic stimulation are in some cases fragmented and incomplete sequences of movements, a result not to be expected if the emotional reactions had been stimulated by a preceding or accompanying emotional sensation similar to those normally evoked by certain external stimuli. Further support comes from studies showing that fear and rage reactions produced by hypothalamic stimulation are not subject to avoidance conditioning, a task which is easily accomplished for normal emotional behavior aroused via the external sensory receptors (Masserman, 1941, 1942, 1943; Brady, 1957; Roberts, 1958a, b; Miller, 1961). This finding indicates that the fear or rage emotional sensations may not be perceived by the animal. Complete isolation of the hypothalamus from all the higher structures of the brain, leaving it attached only to the brain stem, results in a preparation capable of complex, but nondirected, rage behavior (Bard, 1928; Bard and Macht, 1958). The hypothalamic isolation apparently removes the possibility for complex sensory perception but leaves intact the motor mechanisms for a complex behavioral reaction. Finally, to support the view that the hypothalamus is an effector organ independent of the sensors in the emotion mechanisms, there are reports that stimulation in some regions of the hypothalamus in man does not produce any subjective emotional experiences (White, 1940).

There are exceptions to the above studies, however. Some animals show hypothalamically elicited behavior that is well integrated and complete (Hess, 1954). Other studies show that some electrically produced emotional reactions can be conditioned (Miller, 1961). Another exception is that stimulation in the lateral hypothalamic regions in man is reported as pleasant (Heath, 1964). But these studies do not negate the former ones, because some parts of the hypothalamus may actually be sensors in the emotion mechanisms instead of effectors; the point is that the effectors are in the hypothalamus and do not seem to be located in other regions of the brain. Stimulation studies elsewhere in the brain show that perception of emotional sensations always accompanies or precedes the related motor reactions.

Various parts of the brain which when stimulated elicit emotional reactions, or when lesioned produce interferences in the association of emotional sensations with other perceptions, all seem to be parts of a rim (limbus) of cortical and telencephalic structures which surround the diencephalon and basal ganglia. Broca (1878) named the circle of structures the limbic lobe, and Pribram and Kruger (1954) found most of this tissue to constitute the tertiary olfactory system, only three synapses removed from the primary olfactory one. The earlier work of Smith (1897) indicated that this limbic tissue does not underlie olfactory function, as was previously speculated because of its close anatomical proximity to the primary olfactory structures. Papez (1937, 1939) later proposed that this limbic tissue is related to the mechanisms of emotion, although his hypothesis was based on meager clinical evidence. He proposed a circuit of anatomical structures forming a complete circle of connections, which, starting with the lowest diencephalic component, consists of: mammillary bodies, mammillothalamic tract, anterior nucleus of the thalamus, radiations of anterior nucleus of the thalamus, cingulate cortex, cingulum bundle, entorhinal or pyriform temporal cortex, temporoammonic tract, hippocampus, fornix and, finally, back to the mammillary bodies. Other structures are closely interconnected with components of Papez' circular pathway, especially the septal nucleus and the amygdala, and together all these structures constitute what is today commonly called the limbic system. The primary olfactory system is no longer included in the limbic system originally described by Broca. The fact that interruption of Papez' circuit at any one of its many points does not produce a consistent result indicates that the circular integrity of this system is of no particular importance, but the circuitry does point out the high degree of interconnectivity of the elements in the limbic system.

Direct electrical stimulation of the human temporal cortex, cingulate gyrus or septal region can produce conscious emotional sensations of fear, anger, sexual-like pleasure and euphoria (Penfield and Jasper, 1954; Penfield, 1958; Heath, 1964), a finding that indicates that these anatomical structures are sensors of the emotion mechanisms. Lesion of these structures as well as of the hippocampus appears to produce a deficit in the formation of certain types of new associations. In man, lesions of the hippocampus produce a loss of immediate memory (Milner and Penfield, 1955). However, the fact that memories of associations formed before the lesions are not affected indicates that the deficit is a failure to form new associations, not a failure to recall established memories. Lesions of the hippocampus in animals produce deficits in forming associations between neutral stimuli presented separately in temporal succession (successive discrimination), in discovering the correct path through a maze (maze learning) and in learning to avoid a desirable object whose approach brings on punishment (passive avoidance). However, these animals are still capable of learning and remembering the associations formed in a brightness discrimination task (Kimble, 1963), a result which indicates that the deficit is a rather specific associative disorder that does not affect all types of learning and short-term memories. Deficits in performing passive-avoidance tasks are also produced by lesions in the cingulate cortex and septal nuclei (Isaacson and Wickelgren, 1962), although lesions in the former are more effective in producing deficits in active-avoidance tasks in which the subject must learn to form the association of avoiding punishment by reacting to a neutral stimulus (McCleary, 1961).

Lesion of the temporal lobe in animals, which includes part of the hippocampus and temporal cortex and all the amygdala, was first reported to produce an inability to discriminate between edible and inedible objects, loss of fear and aggression, and an inability to distinguish appropriate sexual objects (Kluver and Bucy, 1937; Schreiner and Kling, 1953). Later work indicated that seemingly opposite results could be obtained with the same type of lesion (Bard and Mountcastle, 1947). A resolution between the divergent findings has been provided by showing that factors such as social dominance determine the particular manifestation produced by the lesions. Pribram (1962) found that aggressive animals which had hostile interactions with other aggressive animals before amygdalectomy became docile after the surgery, whereas aggressive animals which had peaceful interactions with docile animals became even more aggressive after the identical ablations. This finding shows the degree of complexity of the previously formed associations which interact with the emotion mechanisms.

It is difficult to make a general statement about the functions of the limbic and hypothalamic mechanisms, although as systems they seem to be involved in the regulatory mechanisms for motivation, for the formation of certain relevant associations and for the performance of the many approach and avoidance behaviors that are necessary for the survival of the animal and its species. Some of the sensory components seem to have emerged from the primary olfactory mechanism (limbic system), and the motor components are closely related to the autonomic effector mechanisms located in the hypothalamus. Direct stimulation in the limbic system produces sensations of emotional feelings, whereas direct stimulation in the hypothalamus seems to produce complex behavioral reactions which may be independent of the relevant emotional sensations. In animals electrical stimulation in either of the two systems will produce approach and avoidance behavioral reactions which may result from elicited emotional sensations or be produced directly by a purely effector mechanism (Olds, 1962; Olds and Olds, 1963). It has been suggested that these brain mechanisms produce directly both the motivation and reinforcement for the learning and performance of complex behaviors (Deutsch and Howarth, 1963), but our knowledge of the nature and operation of such mechanisms is still tenuous and vague.

DISSECTION GUIDE

PREPARATION OF THE BRAIN FOR DISSECTION

The cow and sheep are both members of the ungulate group of animals and have brains that are similar in appearance. In contrast to the smaller brains of the rat and cat, the cow and sheep brains are large enough to dissect in detail and are easy to study with the unaided eye. The cow brain can be obtained from a local meat market, and either of the brains can be ordered from the Carolina Biological Supply Company (Burlington, North Carolina 27215). Because it is larger and more often available locally, the cow brain is perhaps better for learning mammalian neuroanatomy, but since the sheep brain is identical to it in every detail except size, the same guide can be used for the dissection of both.

To prepare a fresh brain for dissection, soak it in a solution of 10 per cent formaldehyde for several days, and then rinse it in water for one day. The pieces of tissue should be kept moist at all times during dissection and stored in a weak 1 per cent solution of formaldehyde. Pieces of cheesecloth can be wrapped around the slices after they have been cut to keep them arranged in serial order during storage.

CORTICAL SURFACE

First examine the whole brain from the ventral, dorsal and lateral views. The cerebellum and medulla are not shown in this dissection guide and are usually absent in brains extracted by meat packers. (Refer to the rat atlas given later for the basic structure of these regions.)

In all views observe that the cortex is highly convoluted or gyrencephalic. A gyrencephalic brain has a greater cortical surface than a smooth or lissencephalic brain of the same weight, and it is often thought that a highly convoluted cortex characterizes a more encephalized or advanced brain. On the other hand, by examining the brains of closely related animals of varying physical sizes, it becomes apparent that brain size is also related to the amount of cortical gyrencephalization (rule of Baillarger and Dareste; see Ariëns Kappers et al., 1960). Apparently both factors are important in determining the degree of convolution of the cortical surface.

There are several important sulci dividing the cortex into separate major regions. Each region seems to underlie a different higher nervous function. The ansate sulcus, seen in the dorsal and lateral views of the ungulate brain, is the homologue of the central sulcus in man, which separates cutaneous sensory cortex from motor cortex. However, in lower animals the cortex surrounding the ansate sulcus has overlapping projections from both sensory and motor systems, giving rise to a combined sensorimotor cortex (Ariëns Kappers et al., 1960). In rats this sensorimotor cortex is overlapped further with projections from systems involving the frontal cortex. The lateral sulcus separates the visual cortex from the more lateral parietal association cortex. The pseudosylvian sulcus is the homologue of the lateral or sylvian sulcus in man, which separates the temporal lobe from the frontal lobe. The suprasylvian sulcus is a long sulcus in lower animals which separates parietal cortex from temporal cortex. The rhinal sulcus seen in the ventral view is more appropriately named the rhinal fissure, because it appears as the only indentation in lissencephalic brains (see the rat brain) and separates the tissue of the limbic system, containing the amygdala and hippocampus, from the more dorsal temporal cortex. The cingulate sulcus seen in the medial view of section X (Fig. 2–15) separates the cingulate or limbic cortex from the more dorsal cutaneous sensory cortex.

The frontal granular cortex in man constitutes approximately one third of the cortical mass and is the area that differentiates most clearly the brain of man from that of the lower animals. The ungulate brain has a small frontal lobe (Ariëns Kappers et al., 1960) which is approximately the same small proportion of the cortical mass as that in the carnivore brain. The frontal cortex anterior and lateral to the coronal sulcus (seen in the lateral view, Fig. 2-3) seems to be anatomically distinct from the sensorimotor cortex in the ungulate.

CORONAL SECTIONS

First cut the whole brain in half by slicing it carefully down the exact middle. Cut the left hemisphere into slabs in the frontal or coronal plane and the right hemisphere into slabs in the lateral or sagittal plane. The orientation of the slices in the left hemisphere is shown in Figure 2-4. The bold line (C) on this plate is the key slice, and the rest are parallel to it. The slice is cut through the anterior commissure and posterior part of the optic chiasm (see section W, Fig. 2-14, for the locations of these and other medial surface structures), which places the anterior and posterior commissures in approximately the same horizontal plane, a convention used so that the ungulate brain can be compared with other mammalian brains described in similar atlases or dissection guides. The cut at section A is made just rostral to the corpus callosum, and the B slice is made halfway between A and C. The D and E slices are made to trisect the massa intermedia of the thalamus. The F section is made just rostral to the superior colliculus, and the G section is made through its center.

The brain slabs caudal to section F will be in two pieces because in this region there are no connective elements between the subcortical and surrounding cortical structures. The subcortical pieces can be fitted together and observed from dorsal and lateral views, exhibiting the geniculate bodies and the colliculi which together form the corpora quadrigemina (see Fig. 2-12). The superior and inferior colliculi together are known as the tectum, and the tissue beneath these structures is the tegmentum.

SAGITTAL SECTIONS

Figure 2-13 shows that section W is through the center of the brain and reveals the medial surface. Section X is made in the right hemisphere close to section W in order to make a thin but whole slab. This thin slice will reveal, on section X, the three distinct fiber tracts in the thalamus: the descending or ventral fornix, the mammillothalamic tract (also known as the tract of Vicq d'Azyr), and the habenulopeduncular tract (also known as tractus retroflexus). Section Y is made by cutting a thin slab which bisects the superior colliculus. Section Z is made just lateral to the colliculi and reveals the cerebral peduncles and internal capsule.

Figures 2–1 to 2–29 follow.

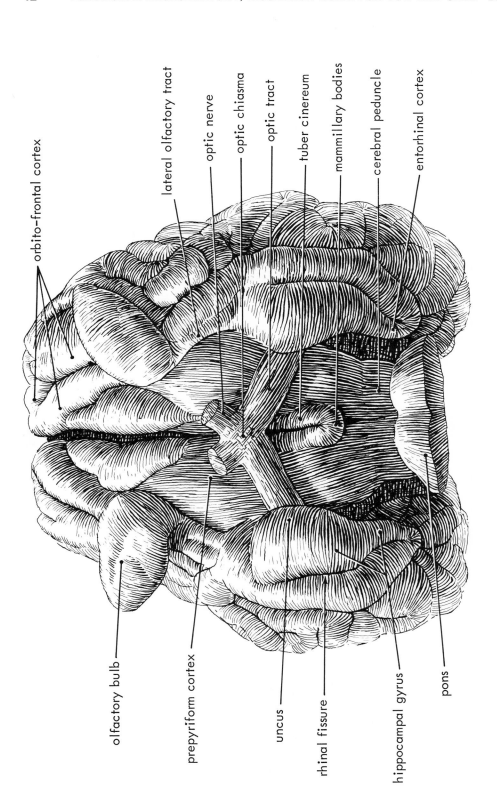

orbito-frontal cortex

lateral olfactory tract

optic nerve

optic chiasma

optic tract

tuber cinereum

mammillary bodies

cerebral peduncle

entorhinal cortex

VENTRAL VIEW

olfactory bulb

prepyriform cortex

uncus

rhinal fissure

hippocampal gyrus

pons

Figure 2–1

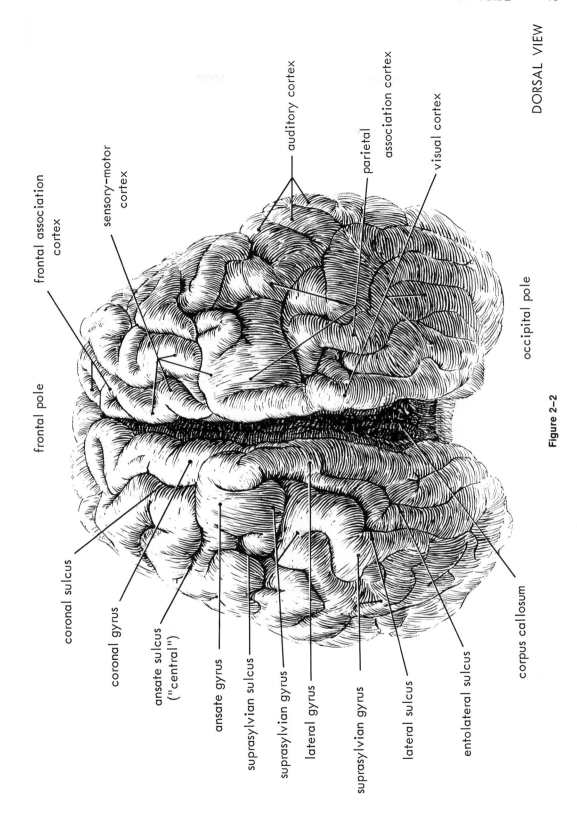

DORSAL VIEW

auditory cortex

parietal association cortex

visual cortex

sensory-motor cortex

frontal association cortex

occipital pole

frontal pole

coronal sulcus

coronal gyrus

ansate sulcus ("central")

ansate gyrus

suprasylvian sulcus

suprasylvian gyrus

lateral gyrus

suprasylvian gyrus

lateral sulcus

entolateral sulcus

corpus callosum

Figure 2–2

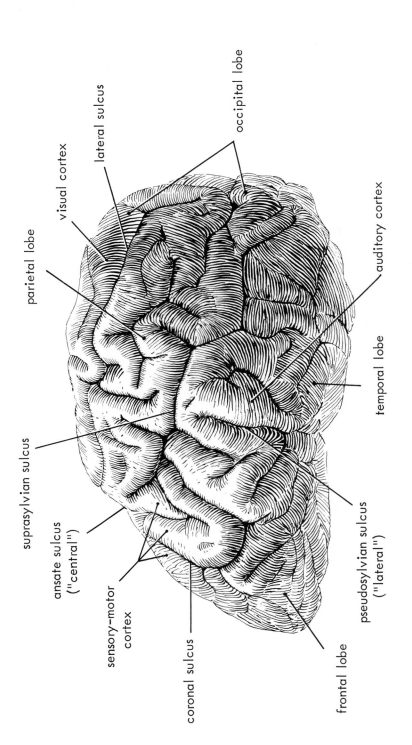

LATERAL VIEW

occipital lobe

lateral sulcus

visual cortex

parietal lobe

auditory cortex

temporal lobe

suprasylvian sulcus

ansate sulcus ("central")

sensory-motor cortex

pseudosylvian sulcus ("lateral")

coronal sulcus

frontal lobe

Figure 2-3

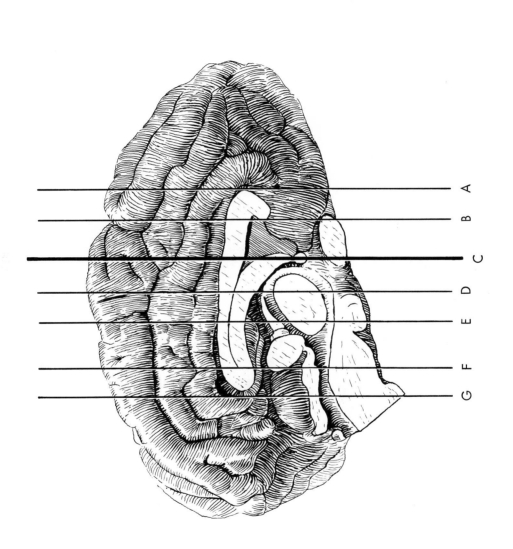

PLANES OF CORONAL SECTIONS

A B C D E F G

Figure 2-4

SECTION A

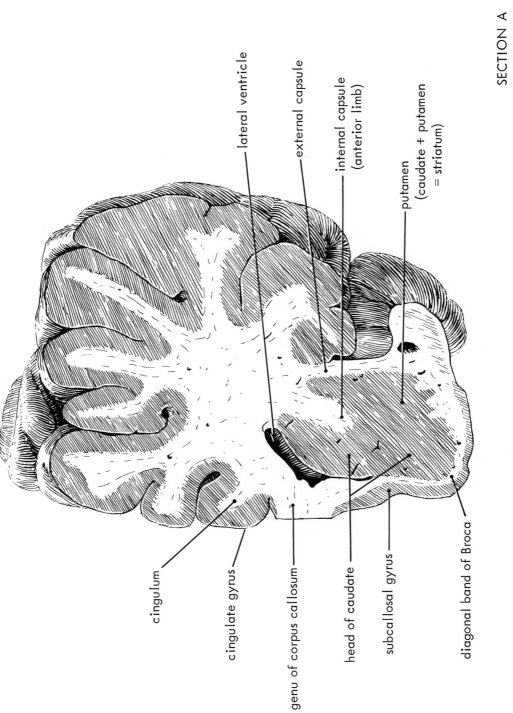

lateral ventricle

external capsule

internal capsule
(anterior limb)

putamen
(caudate + putamen
= striatum)

cingulum

cingulate gyrus

genu of corpus callosum

head of caudate

subcallosal gyrus

diagonal band of Broca

Figure 2–5

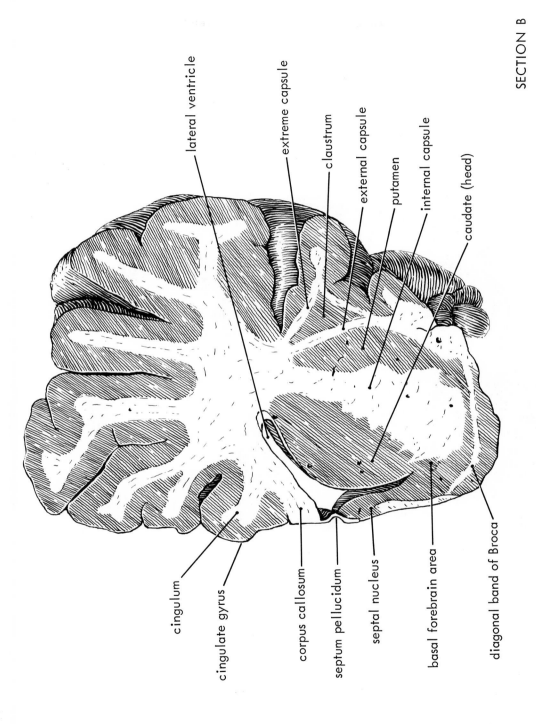

lateral ventricle

extreme capsule

claustrum

external capsule

putamen

internal capsule

caudate (head)

cingulum

cingulate gyrus

corpus callosum

septum pellucidum

septal nucleus

basal forebrain area

diagonal band of Broca

SECTION B

Figure 2–6

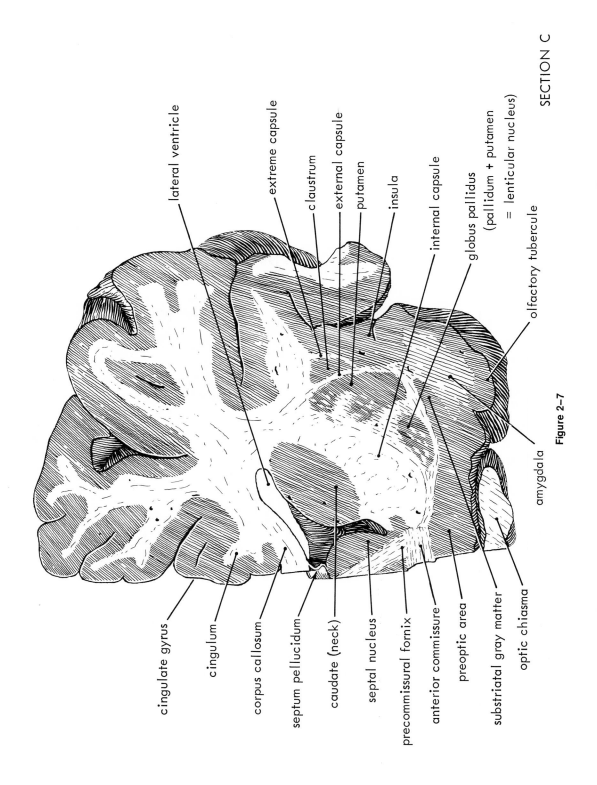

SECTION C

lateral ventricle

extreme capsule

claustrum

external capsule

putamen

insula

internal capsule

globus pallidus
(pallidum + putamen
= lenticular nucleus)

olfactory tubercule

amygdala

cingulate gyrus

cingulum

corpus callosum

septum pellucidum

caudate (neck)

septal nucleus

precommissural fornix

anterior commissure

preoptic area

substriatal gray matter

optic chiasma

Figure 2–7

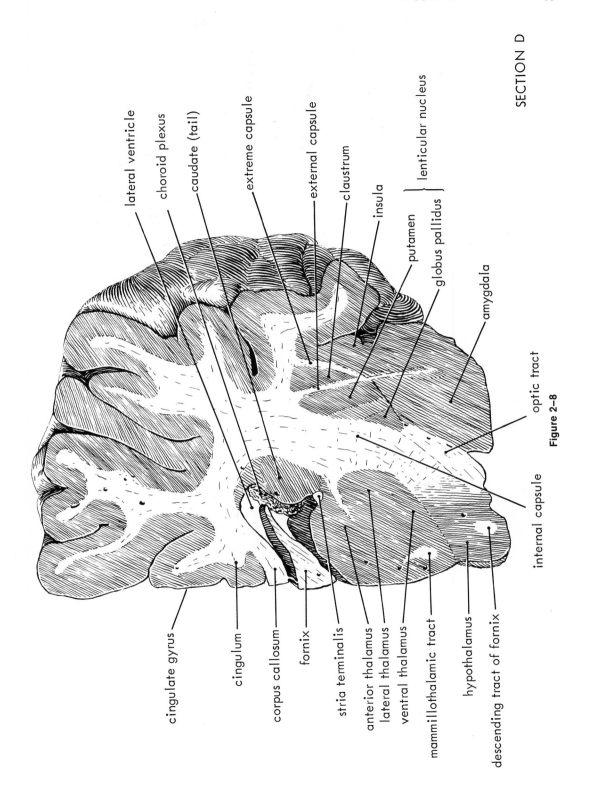

SECTION D

lateral ventricle

choroid plexus

caudate (tail)

extreme capsule

external capsule

claustrum

insula

putamen

globus pallidus

lenticular nucleus

amygdala

optic tract

internal capsule

descending tract of fornix

hypothalamus

mammillothalamic tract

ventral thalamus

lateral thalamus

anterior thalamus

stria terminalis

fornix

corpus callosum

cingulum

cingulate gyrus

Figure 2–8

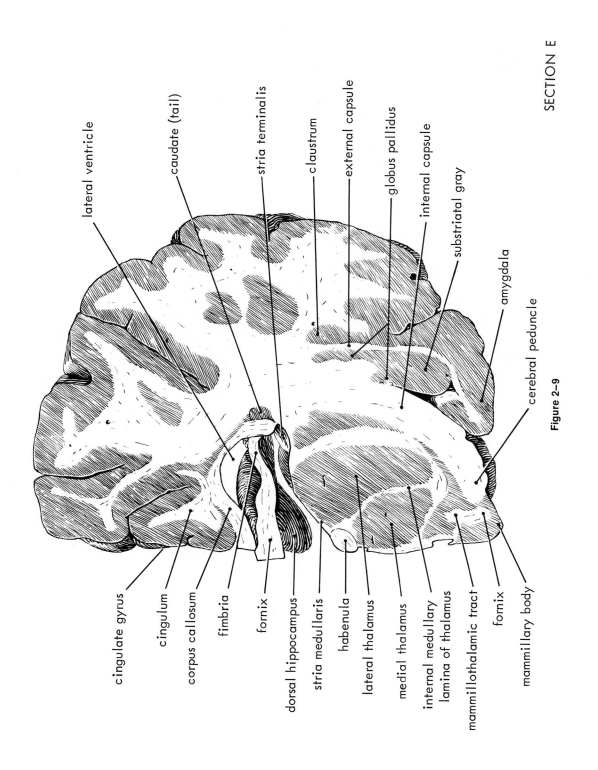

SECTION E

Figure 2-9

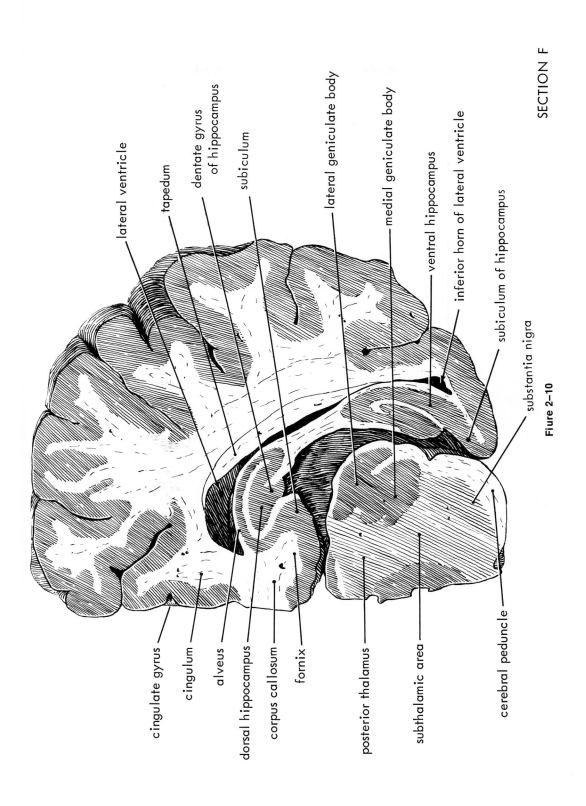

SECTION F

lateral ventricle

tapedum

dentate gyrus
of hippocampus

subiculum

lateral geniculate body

medial geniculate body

ventral hippocampus

inferior horn of lateral ventricle

subiculum of hippocampus

substantia nigra

Fiure 2–10

cingulate gyrus

cingulum

alveus

dorsal hippocampus

corpus callosum

fornix

posterior thalamus

subthalamic area

cerebral peduncle

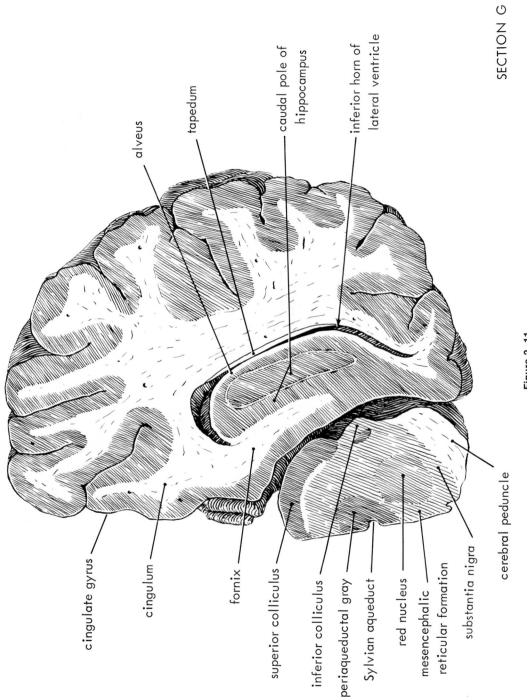

SECTION G

Figure 2–11

cingulate gyrus

cingulum

fornix

superior colliculus

inferior colliculus

periaqueductal gray

Sylvian aqueduct

red nucleus

mesencephalic
reticular formation

substantia nigra

cerebral peduncle

alveus

tapedum

caudal pole of
hippocampus

inferior horn of
lateral ventricle

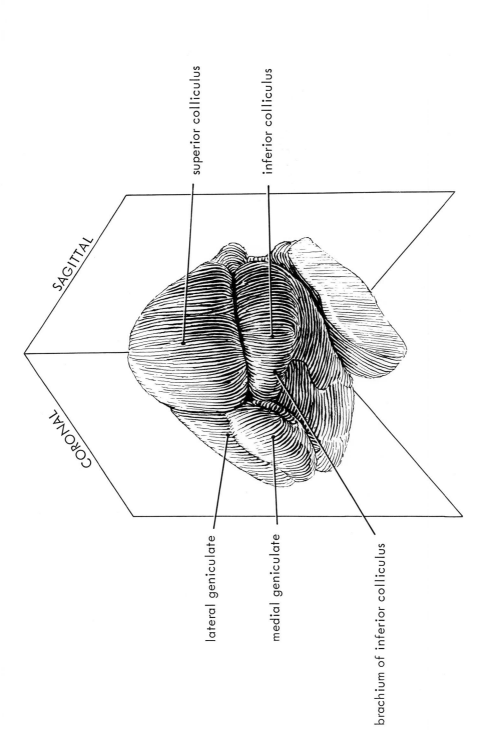

superior colliculus

inferior colliculus

SAGITTAL

CORONAL

lateral geniculate

medial geniculate

brachium of inferior colliculus

CORPORA QUADRIGEMINA

Figure 2–12

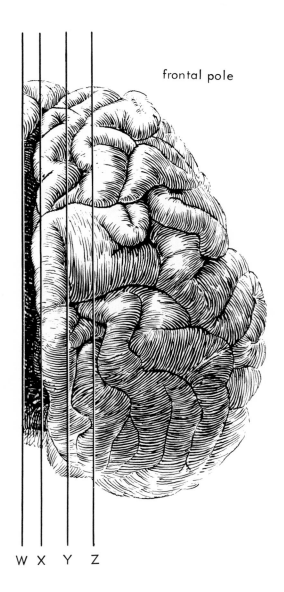

frontal pole

W X Y Z

PLANES OF SAGITTAL SECTIONS
Figure 2–13

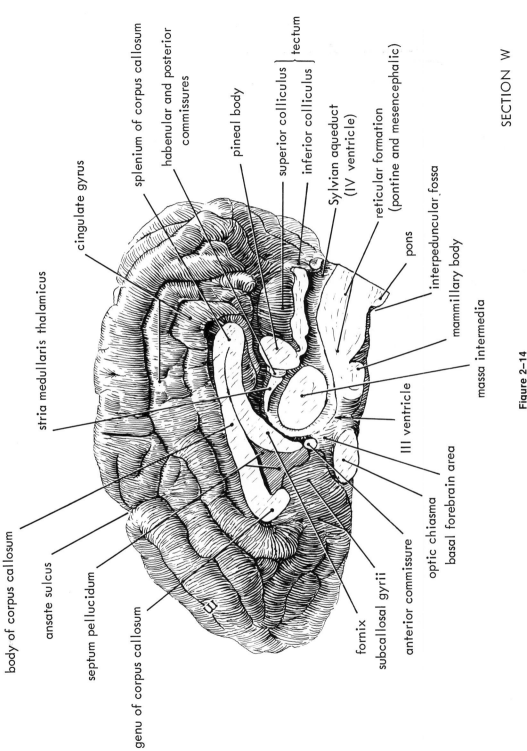

body of corpus callosum

ansate sulcus

septum pellucidum

genu of corpus callosum

stria medullaris thalamicus

cingulate gyrus

splenium of corpus callosum

habenular and posterior commissures

pineal body

superior colliculus ⎫ tectum
inferior colliculus ⎭

Sylvian aqueduct (IV ventricle)

reticular formation (pontine and mesencephalic)

pons

interpeduncular fossa

mammillary body

massa intermedia

III ventricle

basal forebrain area

optic chiasma

anterior commissure

subcallosal gyrii

fornix

SECTION W

Figure 2-14

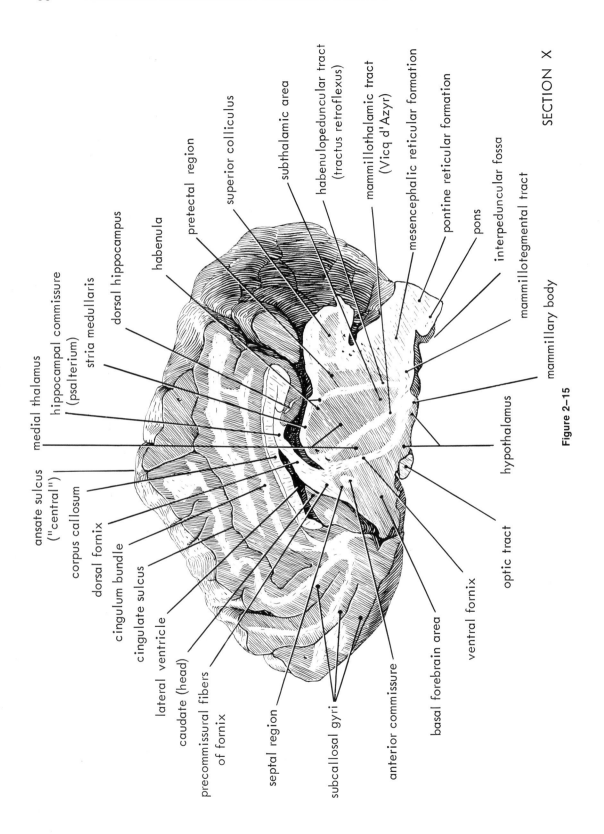

SECTION X

Figure 2-15

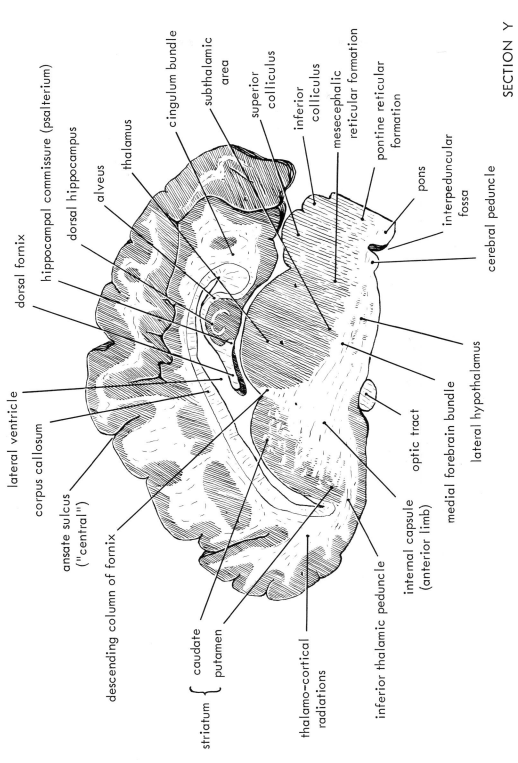

SECTION Y

Figure 2–16

lateral ventricle

corpus callosum

dorsal fornix

hippocampal commissure (psalterium)

dorsal hippocampus

alveus

thalamus

cingulum bundle

subthalamic area

superior colliculus

inferior colliculus

mesecephalic reticular formation

pontine reticular formation

pons

interpeduncular fossa

cerebral peduncle

ansate sulcus ("central")

descending column of fornix

striatum { caudate putamen

thalamo-cortical radiations

inferior thalamic peduncle

internal capsule (anterior limb)

medial forebrain bundle

optic tract

lateral hypothalamus

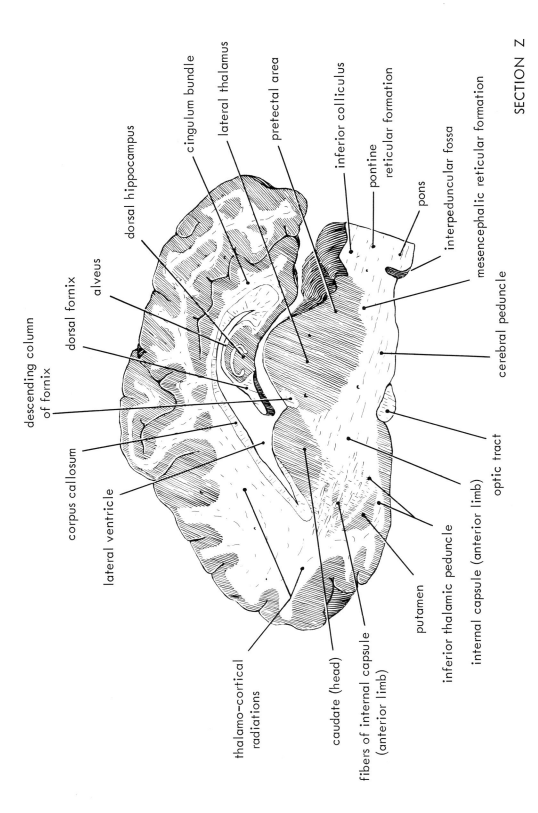

SECTION Z

Figure 2–17

Figures 2-18 and 2-19.   Higher nervous systems of cow and sheep brain. The top two figures show the lateral view (upper left) and mesial view (upper right) of the surfaces of the cerebrum of the cow or sheep brain. The brains of these two animals are identical except for size. The lettered lines shown in the lateral view indicate the planes of section of the coronal slabs. These slabs, lettered A through G, and the sagittal slabs, lettered X through Z, are identical to those shown in the dissection guide. The color legends at the bottom of the pages identify the various systems.

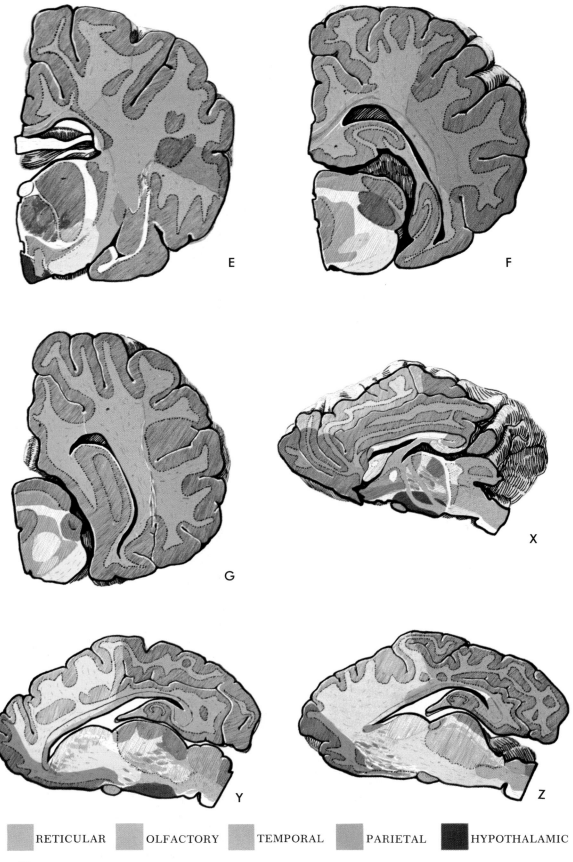

E

F

G

X

Y

Z

RETICULAR OLFACTORY TEMPORAL PARIETAL HYPOTHALAMIC

Figure 2–18

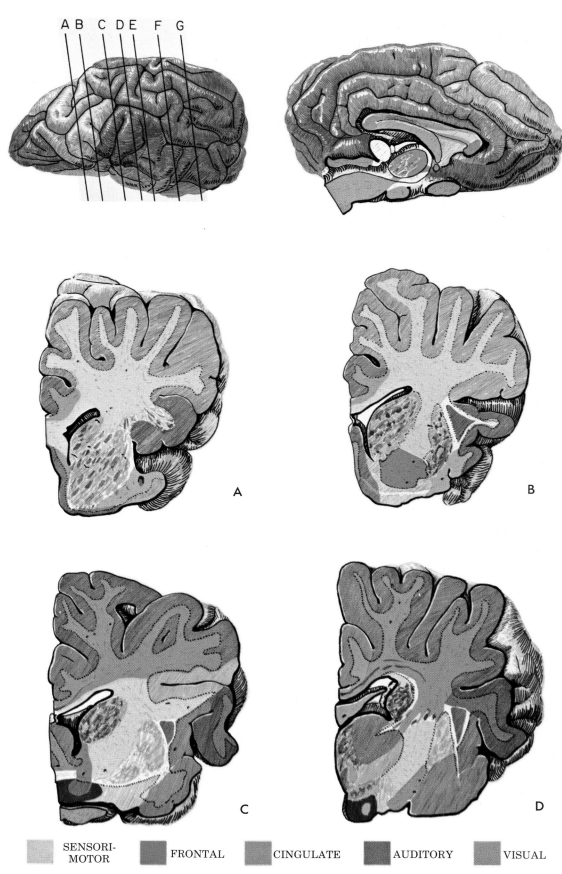

A B C D E F G

SENSORI-MOTOR   FRONTAL   CINGULATE   AUDITORY   VISUAL

A

B

C

D

Figure 2–19

61

Figure 2-20.   Color legend for the various systems of the rat brain. The Broadman numbers of the cytoarchitectonic areas of the cortex are enclosed in parentheses beside each cortical region of the cerebrum. The names of these cortical regions as well as the location of the Broadman areas are adopted from Krieg's collected works (1954, Thomas, Springfield). Each region is identified by a more common name, enclosed in parentheses, as follows: cingulate (anterior limbic); frontal (frontal association and nonspecific thalamocortical); sensorimotor (cutaneous sensory and motor); pyriform (primary olfactory); retrohippocampal (posterior limbic); parietal (parietal association); occipital (primary visual); insular (primary auditory); temporal (temporal association). The olfactory, hypothalamic, and reticular formation systems are seen only in the serial sections of Figures 2-22 and 2-23. The anterior, posterior, and flocculondular lobes of the cerebellum are seen, for the most part, in Figure 2-21.

Figure 2-21.   Cortical components of the various neural systems in the cerebrum and cerebellum of the rat brain. See Figure 2-20 to identify colored systems and abbreviations. Upper figure, lateral view; middle figure, dorsal view; lower figure, mesial view.

CEREBRUM

CINGULATE (23, 24, 25, 25a, 29b, 29c, 32)

FRONTAL—SENSORIMOTOR (4, 6, 8, 8a, 10, 11)

PYRIFORM (51a, 51b, 51e, 51f, 51g, 51h, Tol)

RETROHIPPOCAMPAL (27, 28a, 35, 49, Amm, Fd, Sub)

PARIETAL (1, 2, 2a, 3, 7, 39, 40)

OCCIPITAL (17, 18, 18a, 36)   OLFACTORY (serial sections only)

INSULAR (13, 14)   HYPOTHALAMIC   (″)

TEMPORAL (20, 41)   RETICULAR FORMATION (″)

CEREBELLUM

LINGULA (L)

CENTRALIS (Ce)   Anterior Lobe

CULMEN (Cu)

FISSURA PRIMA (FP)

SIMPLEX (S)

DECLIVE (D)

ANSIFORMIS (ANS)   Posterior Lobe

PARAMEDIANUS (PM)

TUBER (T)

FISSURA PREPYRAMIDALIS (FPP)

PYRAMIS (P)

UVULA (U)

FISSURA POSTEROLATERALIS (FPL)

NODULUS (N)

PARAFLOCCULUS (PF)   Flocculonodular Lobe

Figure 2—20

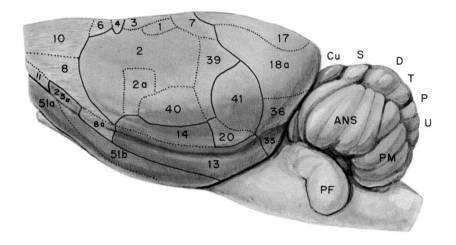

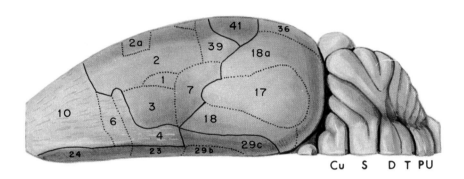

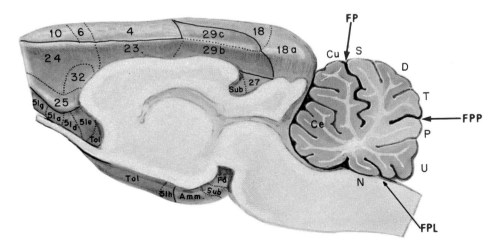

Figure 2–21

Figures 2-22 and 2-23. Subcortical components of the various neural systems in the cerebrum of the rat brain. These coronal sections are the same as those labeled and shown in the stereotaxic atlas of the rat brain. See Figure 2-20 to identify the colored systems.

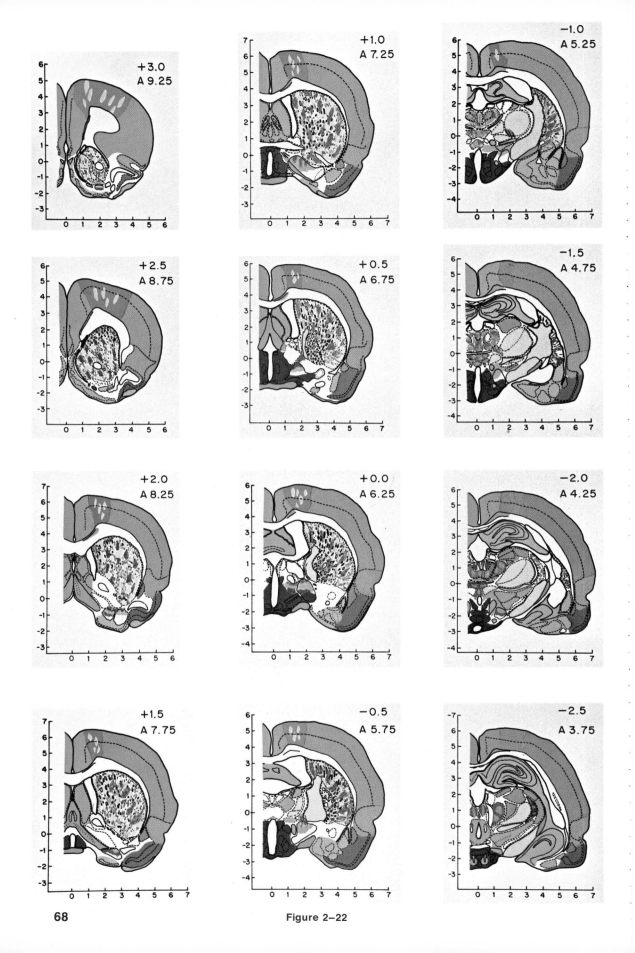

Figure 2–22

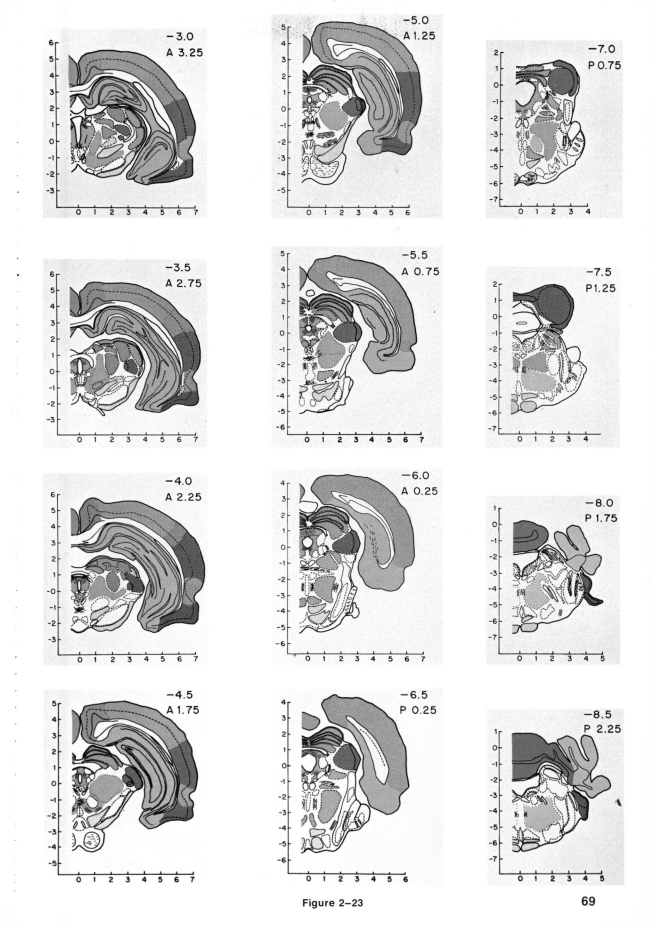

Figure 2–23

Figures 2–24 through 2–29 (pages 71–82) are provided for the use of the reader. They can be removed from the book by tearing them carefully along the inner margin.

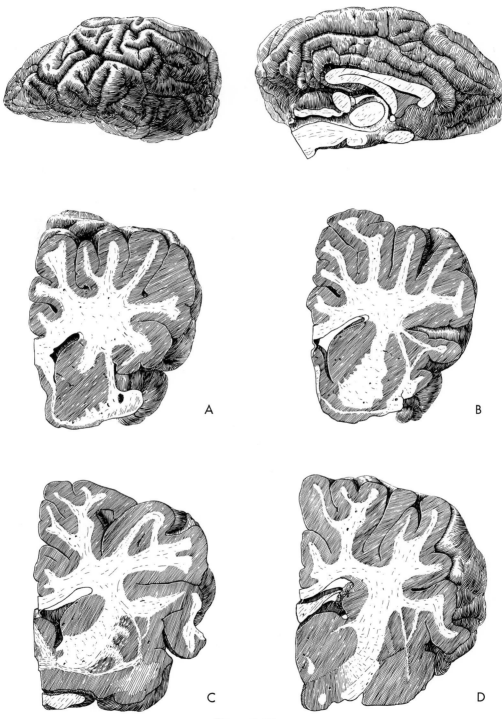

Figure 2–24

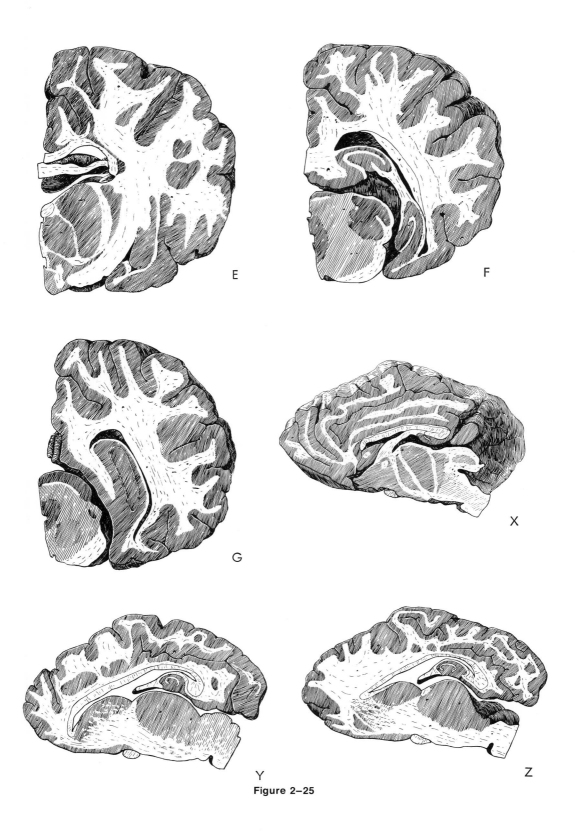

E

F

G

X

Y

Z

Figure 2–25

CEREBRUM

☐ CINGULATE (23, 24, 25, 25a, 29b, 29c, 32)

☐ FRONTAL–SENSORIMOTOR (4, 6, 8, 8a, 10, 11)

☐ PYRIFORM (51a, 51b, 51e, 51f, 51g, 51h, Tol)

☐ RETROHIPPOCAMPAL (27, 28a, 35, 49, Amm, Fd, Sub)

☐ PARIETAL (1, 2, 2a, 3, 7, 39, 40)

☐ OCCIPITAL (17, 18, 18a, 36)   ☐ OLFACTORY (serial sections only)

☐ INSULAR (13, 14)   ☐ HYPOTHALAMIC   (″)

☐ TEMPORAL (20, 41)   ☐ RETICULAR FORMATION (″)

CEREBELLUM

☐ LINGULA (L)

☐ CENTRALIS (Ce)   } Anterior Lobe

☐ CULMEN (Cu)

FISSURA PRIMA (FP)

☐ SIMPLEX (S)

☐ DECLIVE (D)

☐ ANSIFORMIS (ANS)   } Posterior Lobe

☐ PARAMEDIANUS (PM)

☐ TUBER (T)

FISSURA PREPYRAMIDALIS (FPP)

☐ PYRAMIS (P)

☐ UVULA (U)

FISSURA POSTEROLATERALIS (FPL)

☐ NODULUS (N)   } Flocculonodular Lobe

☐ PARAFLOCCULUS (PF)

Figure 2–26

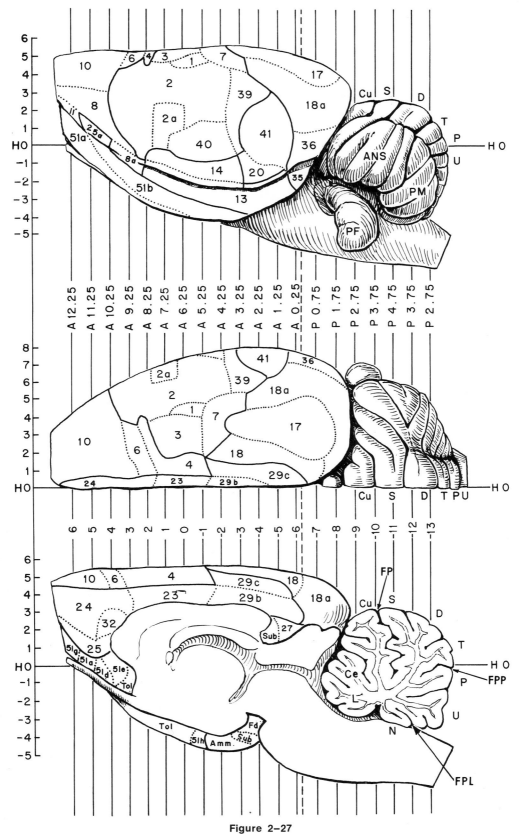

Figure 2–27

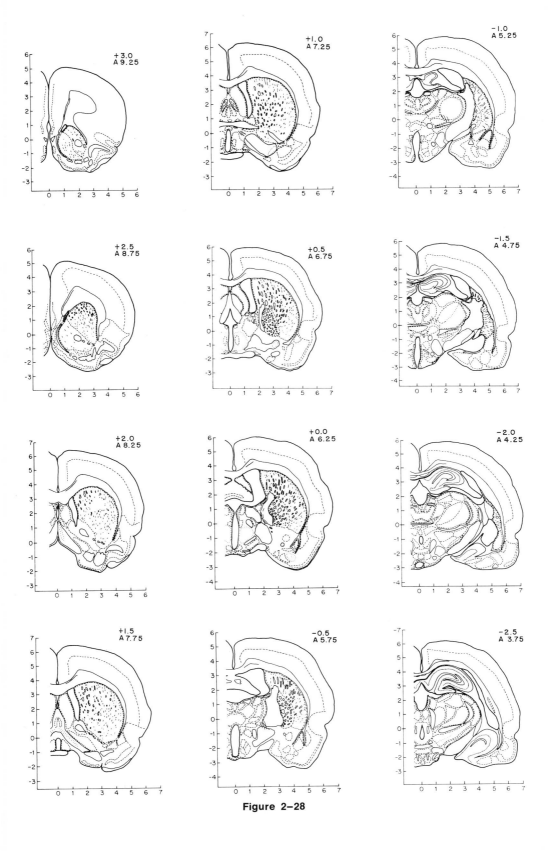

Figure 2–28

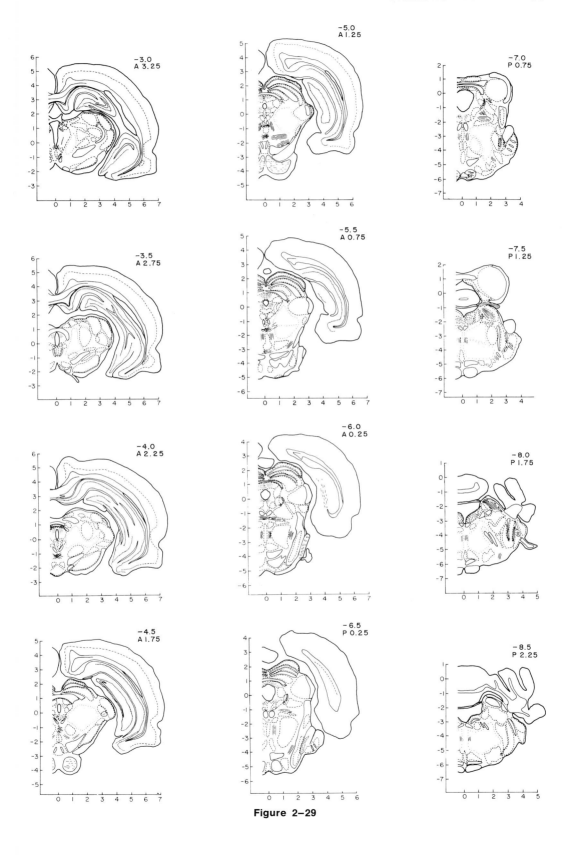

Figure 2–29

REFERENCES

Adrian, E. D. (1944). Brain rhythms. *Nature Rondon. 153*:360–362.

Akert, K. (1961). Diencephalon. *In* Sheer, D. E. (ed.): *Electrical Stimulation of the Brain.* Austin, University of Texas Press, pp. 288-310.

Akert, K., Koella, W. P., and Hess, R., Jr. (1952). Sleep produced by electrical stimulation of the thalamus. *Amer. J. Physiol. 168*:260-267.

Akimoto, H., Yamaguchi, N., Okabe, K., Nakagawa, T., Nakumura, I., Abe, K., Torri, H., and Masahashi, K. (1956). On sleep induced through electrical stimulation of dog thalamus. *Folia Psychiat. Neurol. Jap. 10*:117-146.

Andersson, B. (1951). The effect and localization of electrical stimulation of certain parts of the brain stem in sheep and goats. *Acta Physiol. Scand. 23*:8-24.

Andersson, B., Jewell, P. A., and Larsson, S. (1958). An appraisal of the effects of diencephalic stimulation of conscious animals in terms of normal behavior. *In* Wolstenholme, G. E., and O'Connor, V. M. (eds.): *Neurological Basis of Behavior.* London, J. & A. Churchill, Ltd., pp. 76-89.

Ariëns Kappers, C. U., Huber, G. C., and Crosby, E. C. (1960). *The Comparative Anatomy of the Nervous System of Vertebrates, Including Man.* New York, Hafner Publishing Co.

Bard, P. A. (1928). A diencephalic mechanism for the expression of rage with special reference to the sympathetic nervous system. *Amer. J. Physiol. 84*:490-515.

Bard, P., and Macht, M. B. (1958). The behavior of chronically decerebrate cats. *In* Wolstenholme, G. E., and O'Connor, V. M. (eds.): *Neurological Basis of Behavior.* London, J. & A. Churchill, Ltd., pp. 55-71.

Bard, P., and Mountcastle, V. B. (1947). Some forebrain mechanisms involved in expression of angry behavior. *Ass. Res. Nerv. Dis. Proc. 27*:362-404.

Brady, J. V. (1957). A comparative approach to the experimental analysis of emotional behavior. *In* Hoch, P. H., and Zubin, J. (eds.): *Experimental Psychopathology.* New York, Grune & Stratton.

Brady, J. V. (1958). Ulcers in executive monkeys. *Sci. Amer. 199*:95-100.

Bremer, F., and Stoupel, N. (1959). Facilitation et inhibition des potentiels évoqués corticaux dans l'éveil cérébral. *Arch. Int. Physiol. 67*:1–35.

Broca, P. (1878). Anatomie comparee des circonvolutions cérébrales. Le grand lobe limbique et la scissure limbique dans la série des mammifères. *Rev. Anthropol. 1*:385-498.

Brodal, A. (1957). *The Reticular Formation of the Brain Stem. Anatomical Aspects and Functional Correlations.* Edinburgh, Oliver & Boyd, Ltd.

Brugger, M. (1943). Fresstreib als hypothalamisches symptom. *Helv. Physiol. Pharmacol. Acta 1*:183-198.

Brush, E. S., Mishkin, M., and Rosvold, H. E. (1961). Effects of object preferences and aversions on discrimination learning in monkeys with frontal lesions. *J. Comp. Physiol. Psychol. 54*:319-325.

Buchwald, N. A., Wyers, E. J., Lauprecht, C. W., and Heuser, G. (1961). The "caudate-spindle." IV. A behavioral index of caudate-induced inhibition. *Electroenceph. Clin. Neurophysiol. 13*:531-537.

Buser, P., Rougeul, A., and Perret, C. (1964). Caudate and thalamic influences on conditioned motor responses in the cat. *Bol. Inst. Estud. Med. Biol. (Mex.) 22*:293-307.

Cannon, W. B. (1927). The James-Lange theory of emotions: A critical examination and an alteration. *Amer. J. Psychol. 39*:106-124.

Cannon, W. B. (1931). Again the James-Lange and the thalamic theories of emotion. *Psychol. Rev. 38*:281-295.

Chow, K. L. and Hutt, P. J. (1953). The "association cortex" of *Macaca mulatta*: A review of recent contributions to its anatomy and functions. *Brain 76*:625-677.

Dana, C. L. (1921). The anatomic seat of the emotions: A discussion of the James-Lange theory. *Arch. Neurol. Psychiat. (Chicago) 6*:634-639.

Delgado, J. M. R., and Anand, B. K. (1953). Increased food intake induced by electrical stimulation of the lateral hypothalamus. *Amer. J. Physiol. 172*:162-168.

Dempsey, E. W., and Morison, R. S. (1942a). The production of rhythmically recurrent cortical potentials after localized thalamic stimulation. *Amer. J. Physiol. 135*:293-300.

Dempsey, E. W., and Morison, R. S. (1942b). The interaction of certain spontaneous and induced cortical potentials. *Amer. J. Physiol. 135*:301-308.

Deutsch, J. A., and Howarth, C. I. (1963). Some tests of a theory of intracranial self-stimulation. *Psychol. Rev. 70*:444-460.

Dumont, S., and Dell, P. (1960). Facilitation réticulaire des mécanismes visuels corticaux. *Electroenceph. Clin. Neurophysiol. 12*:769-796.

Fuster, J. M. (1958). Effects of stimulation of brain stem on tachistoscopic perception. *Science 127*:150-151.

Galambos, R., and Sheatz, G. C. (1962). An electroencephalographic study of classical conditioning. *Amer. J. Physiol. 203*:173-184.

Galambos, R., Sheatz, G., and Vernier, V. G. (1956). Electrophysiological correlates of a conditioned response in cats. *Science* 123:376-377.

Garcia-Austt, E. (1963). Influence of the states of awareness upon sensory evoked potentials. *Electroenceph. Clin. Neurophysiol.* Suppl. 24:76-89.

Garcia-Austt, E., Bogacz, J., and Vanzulli, A. (1964). Effects of attention and inattention upon visual evoked responses. *Electroenceph. Clin. Neurophysiol.* 17:136-143.

Haider, M., Spong, P., and Lindsley, D. B. (1964). Attention, vigilance, and cortical evoked-potentials in humans. *Science* 145:180-182.

Hanbery, J., Ajmone Marsan, C., and Dilworth, M. (1954). Pathways of nonspecific thalamo-cortical projection system. *Electroenceph. Clin. Neurophysiol.* 6:103-118.

Hearst, E., Beer, B., Sheatz, G., and Galambos, R. (1960). Some electro-physiological correlates of conditioning in the monkey. *Electroenceph. Clin. Neurophysiol.* 12:137-152.

Heath, R. G. (1964). Pleasure response of human subjects to direct stimulation of the brain: Physiologic and psychodynamic considerations. *In* Heath, R. G. (ed.). *The Role of Pleasure in Behavior.* New York, Hoeber Medical Divison, Harper & Row, pp. 219-250.

Hess, W. R. (1928). Stammgarglien-reizversuche. (Vehr. Deutsch. Physiol. Ges., Sept., 1927). *Ber. Ges. Physiol.* 42:554.

Hess, W. R. (1936). Hypothalamus und die Zentren des autonomen Nerven-systems: Physiologie. *Arch. Psychiat. Nervenkr.* 104:548-557.

Hess, W. R. (1944). Das Schlafsyndrom als Folge dienzephaler Reizung. *Helv. Physiol. Pharmacol. Acta* 2:305-344.

Hess, W. R. (1949). *Das Zeischenhirn: Syndrome, Lokalizationen, Funktionen.* Basel, Schwabe & Co.

Hess, W. R. (1954). *Diencephalon: Autonomic and Extrapyramidal Functions.* New York, Grune & Stratton.

Hinsey, J. C. (1940). The hypothalamus and somatic responses. *Res. Publ. Ass. Res. Nerv. Ment. Dis.* 20:657-688.

Hubel, D. H., and Wiesel, T. N. (1959). Receptive fields of single neurones in the cat's striate cortex. *J. Physiol. (London)* 148:574-591.

Hubel, D. H., and Wiesel, T. N. (1962). Receptive fields, binocular interaction and functional architecture in the cat's visual center. *J. Physiol. (London)* 160:106-154.

Hubel, D. H., and Wiesel, T. N. (1963a). Shape and arrangement of columns in cat's striate cortex. *J. Physiol. (London)* 165:559-568.

Hubel, D. H., and Wiesel, T. N. (1963b). Receptive fields of cells in striate cortex of very young, visually inexperienced kittens. *J. Neurophysiol.* 26:994-1002.

Hubel, D. H., and Wiesel, T. N. (1963c). Single-cell responses in striate cortex of kittens deprived of vision in one eye. *J. Neurophysiol.* 26:1003-1017.

Hubel, D. H., and Wiesel, T. N. (1965). Receptive fields and functional architecture in two nonstriate visual areas (18 and 19) of the cat. *J. Neurophysiol.* 28:229-289.

Hume, D. (1739). *A Treatise of Human Nature.* 1911 edition published by J. M. Bent & Sons, Ltd., London.

Hunsperger, R. W. (1956). Affektreaktionen auf elektrische Reizung in Hirnstamm der Katze. *Helv. Physiol. Pharmacol. Acta* 14:70-92.

Hunter, J., and Jasper, H. H. (1949). Effects of thalamic stimulation in unanaesthetised animals: The arrest reaction and petit mal-like seizures, activation patterns and generalized convulsions. *Electroenceph. Clin. Neurophysiol.* 1:305-324.

Ingram, W. R. (1952). Brain stem mechanisms in behavior. *Electroenceph. Clin. Neurophysiol.* 4:397-406.

Ingram, W. R., Barris, R. W., and Randon, S. W. (1936). Catalepsy: An experimental study. *Arch. Neural. Psychiat. (Chicago)* 35:1175-1197.

Isaacson, R. L., and Wickelgren, W. O. (1962). Hippocampal ablation and passive avoidance. *Science* 138:1104-1106.

James, W. (1890). *The Principles of Psychology.* New York, Henry Holt & Co., Inc.

Jasper, H. H. (1960). Unspecific thalamocortical relations. *In* Field, J., Magoun, H. W., and Hall, V. E. (eds.): *Handbook of Physiology, Section 1: Neurophysiology, Vol. II.* Washington, D.C., American Physiological Society, pp. 1307-1321.

Kant, I. (1781). *Critik der reinen Vernunft.* As translated by M. Müller. New York, The Macmillan Co., 1896.

Karplus, J. P. (1937). Die Physiologie der vegetativen Zentren. Auf Grund experimenteller Erfahrungen. *In* Bumke, O., and Foerster, O. (eds.): *Handbuch der Neurologie.* Berlin, Springer-Verlag.

Kimble, D. P. (1963). The effects of bilateral hippocampal lesions in rats. *J. Comp. Physiol. Psychol.* 56:273-283.

Kluver, H., and Bucy, P. C. (1937). "Psychic blindness" and other symptoms following bilateral temporal lobectomy in rhesus monkeys. *Amer. J. Physiol.* 119:352-353.

Konorski, J., and Lawicka, W. (1964). Analysis of errors by prefrontal animals on the delayed-response test. *In* Warren, J. M., and Akert, K. (eds.): *The Frontal Granular Cortex and Behavior.* New York, McGraw-Hill Book Co., Inc., pp. 271-294.

Lange, C. G. (1885). *Om Sindsbevaegelser. et. psyko. fysiolog. studie.* Copenhagen, Kronar.

Larsson, S. (1954). On the hypothalamic organization of the nervous mechanism regulating food intake. Part I. *Acta Physiol. Scand.* 32:Suppl. 115.

Lettvin, J. Y. (1967). Public lecture given at California Institute of Technology.

Lettvin, J. Y., Maturana, H. R., McCulloch, W. S., and Pitts, W. H. (1959). What the frog's eye tells the frog's brain. *Proc. Inst. Radio Engineer.* 47:1940-1951.

Lindsley, D. B., Bowden, J. W., and Magoun, H. W. (1949). Effect upon the EEG of acute injury to the brain stem activating system. *Electroenceph. Clin. Neurophysiol.* 1:475-486.

Lindsley, D. B., Schreiner, L. H., Knowles, W. B., and Magoun, H. W. (1950). Behavioral and EEG changes following chronic brain stem lesions in the cat. *Electroenceph. Clin. Neurophysiol.* 2:483-498.

Locke, J. (1690). *An Essay Concerning Human Understanding and a Treatise on the Conduct of the Understanding.* 1860 edition published by Hayes & Zell, Philadelphia.

MacLean, P. D. (1949). Psychosomatic disease and the "visceral brain": Recent developments bearing on the Papez theory of emotion. *Psychosom. Med.* 11:338-353.

MacLean, P. D. (1958). The limbic systems with respect to self-preservation and the preservation of the species. *J. Nerv. Ment. Dis.* 127:1-11.

MacLean, P. D., and Ploog, D. W. (1962). Cerebral representation of penile erection. *J. Neurophysiol.* 25:30-55.

Malmo, R. B. (1942). Interference factors in delayed response in monkeys after removal of frontal lobes. *J. Neurophysiol.* 5:295-308.

Marsh, J. T., McCarthy, D. A., Sheatz, G., and Galambos, R. (1961). Amplitude changes in evoked auditory potentials during habituation and conditioning. *Electroenceph. Clin. Neurophysiol.* 13:224-234.

Masserman, J. H. (1941). Is the hypothalamus a center of emotion? *Psychosom. Med.* 5:3-25.

Masserman, J. H. (1942). Hypothalamus in psychiatry. *Amer. J. Psychiat.* 98:633-637.

Masserman, J. H. (1943). *Behavior and Neurosis.* Chicago, University of Chicago Press.

Maturana, H. R., Lettvin, J. Y., McCulloch, W. S., and Pitts, W. H. (1960). Anatomy and physiology of vision in the frog (*Rana pipiens*). *J. Gen. Physiol.* 43:129-176.

McCleary, R. A. (1961). Response specificity in the behavioral effects of limbic system lesions in the cat. *J. Comp. Physiol. Psychol.* 54:605-613.

Miller, N. E. (1961). Learning and performance motivated by direct stimulation of the brain. *In* Sheer, D. E. (ed.): *Electrical Stimulation of the Brain.* Austin, University of Texas Press, pp. 387-396.

Milner, B. (1964). Some effects of frontal lobectomy in man. *In* Warren, J. M., and Akert, K. (eds.): *The Frontal Granular Cortex and Behavior.* New York, McGraw-Hill Book Co., Inc., pp. 313-334.

Milner, B., and Penfield, W. (1955). The effect of hippocampal lesions on recent memory. *Trans. Amer. Neurol. Ass.* 80:42-48.

Mishkin, M. (1964). Perseveration of central sets after frontal lesions in monkeys. *In* Warren, J. M., and Akert, K. (eds.): *The Frontal Granular Cortex and Behavior.* New York, McGraw-Hill Book Co., Inc., pp. 219-241.

Mishkin, M., and Pribram, K. H. (1956). Analysis of the effects of frontal lesions in the monkey. II. Variations of delayed response. *J. Comp. Physiol. Psychol.* 49:36-40.

Mishkin, M., Prockop, E. S., and Rosvold, H. E. (1962). One-trial object-discrimination learning in monkeys with frontal lesions. *J. Comp. Physiol. Psychol.* 55:178-181.

Morison, R. S., and Dempsey, E. W. (1942). A study of thalamocortical relations. *Amer. J. Physiol.* 135:281-292.

Morison, R. S., and Dempsey, E. W. (1943). Mechanism of thalamocortical augmentation and repetition. *Amer. J. Physiol.* 138:297-308.

Morrell, F. (1967). Electrical signs of sensory coding. *In* Quarton, G. C., Melnechuk, T., and Schmitt, F. O. (eds.): *The Neurosciences.* New York, Rockefeller University Press, pp. 452-468.

Morrell, F., and Jasper, H. H. (1956). Electrographic studies of the formation of temporary connections in the brain. *Electroenceph. Clin. Neurophysiol.* 8:201-215.

Moruzzi, G., and Magoun, H. W. (1949). Brain stem reticular formation and activation of the EEG. *Electroenceph. Clin. Neurophysiol.* 1:455-473.

Müller, J. (1838). *Handbuch der Physiologie des Menschen für Vorlesungen.* Coblenz, Hölscher.

Nakao, H., and Maki, T. (1958). Effect of electrical stimulation of the nucleus caudatus upon conditioned avoidance behavior in the cat. *Folia Psychiat. Neurol. Jap.* 12:258-264.

Narikashvili, S. P. (1963). Influence of unspecific impulses on the responses of sensory cortex. *In* Moruzzi, G., Fessard, A., and Jasper, H. H. (eds.): *Brain Mechanisms. Progress in Brain Research. Vol. 1.* Amsterdam, Elsevier Publishing Co., pp. 155-183.

Nielsen, J. M. (1946). *Agnosia, Apraxia, Aphasia.* New York, Paul B. Hoeber, Inc.

Olds, J. (1958). Self-stimulation of the brain. *Science* 127:315-323.

Olds, J. (1962). Hypothalamic substrates of reward. *Physiol. Rev.* 42:554-604.

Olds, M. E., and Olds, J. (1963). Approach-avoidance analysis of rat diencephalon. *J. Comp. Neurol. 120:259-295.*

Papez, J. W. (1937). A proposed mechanism of emotion. *Arch. Neurol. Psychiat (Chicago)* 3:230-251.

Papez, J. W. (1939). Cerebral mechanisms. *Res. Publ. Ass. Res. Nerv. Ment. Dis.* 89:145-159.

Parkes, C. M. (1967). Bereavement. *Brit. Med. J.* 3:232.

Penfield, W. (1958). *The Excitable Cortex in Conscious Man.* Springfield, Ill., Charles C Thomas.

Penfield, W., and Jasper, H. (1954). *Epilepsy and the Functional Anatomy of the Human Brain.* Boston, Little, Brown & Company.

Pribram, K. H. (1954). Toward a science of neuropsychology (method and data). *In* Patton, R. A. (ed.): *Current Trends in Psychology and the Behavioral Sciences.* Pittsburgh, University of Pittsburgh Press, pp. 115-142.

Pribram, K. H. (1962). Interrelations of psychology and the neurological disciplines. *In* Koch, S. (ed.): *Psychology: A Study of a Science. Vol. 4.* New York, McGraw-Hill Book Co., Inc., pp. 119-157.

Pribram, K. H., and Kruger, L. (1954). Functions of the "olfactory brain." *Ann. N.Y. Acad. Sci.* 58:109-138.

Ranson, S. W. (1939). Somnolence caused by hypothalamic lesions in the monkey. *Arch. Neurol. Psychiat. (Chicago) 41:1-23.*

Ranson, S. W., and Magoun, H. W. (1939). The hypothalamus. *Ergebn. Physiol. 41:56-163.*

Rheinberger, M., and Jasper, H. H. (1937). Electrical activity of the cerebral cortex in the unanesthetized cat. *Amer. J. Physiol.* 119:186-196.

Roberts, W. W. (1958a). Rapid escape learning without avoidance learning motivated by hypothalamic stimulation in cats. *J. Comp. Physiol. Psychol.* 51:391-399.

Roberts, W. W. (1958b). Both rewarding and punishing effects from stimulation of posterior hypothalamus of cats with same electrode at same intensity. *J. Comp. Physiol. Psychol. 51:400-407.*

Rossi, G. F., and Zanchetti, A. (1957). The brain stem reticular formation: Anatomy and physiology. *Arch. Ital. Biol. 95:199-435.*

Satterfield, J. H., and Cheatum, D. (1964). Evoked cortical potential correlates of attention in human subjects. *Electroenceph. Clin. Neurophysiol.* 17:456-457.

Scheibel, M. E., and Scheibel, A. B. (1958). Structural substrates for integrative patterns in the brain stem reticular core. *In* Jasper, H. H., Proctor, L. D., Knighton, R. S., Noshay, W. C., Costello, R. T. (eds.): *Reticular Formation of the Brain.* Boston, Little, Brown & Company.

Schlag, J. D., Kuhn, R. L., and Velasco, M. (1966). An hypothesis on the mechanism of cortical recruiting responses. *Brain Res.* 1:208-212.

Schreiner, L., and Kling, A. (1953). Behaviorial changes following rhinencephalic injury in cat. *J. Neurophysiol.* 16:643-659.

Segundo, J. P., Arana, R., and French, J. D. (1955). Behaviorial arousal by stimulation of the brain in the monkey. *J. Neurosurg.* 12:601-613.

Sherrington, C. S. (1900). Experiments on the value of vascular and visceral factors for the genesis of emotion. *Proc. Roy. Soc. Biol.* 66:390-403.

Skinner, J. E., and Lindsley, D. B. (1967). Electrophysiological and behaviorial effects of blockade of the nonspecific thalamocortical system. *Brain Res.* 6:95-118.

Skinner, J. E., and Lindsley, D. B. (1971). Enhancement of visual and auditory evoked potentials during blockade of the non-specific thalamocortical system. *Electroenceph. Clin. Neurophysiol.* In press.

Smith, G. E. (1897). The fornix superior. *J. Anat. Physiol.* 31:80-94.

Smith, O. A., Jr. (1961). Food intake and hypothalamic stimulation. *In* Sheer, D. E. (ed.): *Electrical Stimulation of the Brain.* Austin, University of Texas Press, pp. 367-370.

Spong, P., Haider, M., and Lindsley, D. B. (1965). Selective attentiveness and cortical evoked responses to visual and auditory stimuli. *Science* 148:395-397.

Sterman, M. B., and Clemente, C. D. (1961). Cortical recruitment and behavioral sleep induced by basal forebrain stimulation. *Fed. Proc.* 20:334.

Teuber, H. L. (1964). The riddle of frontal lobe function in man. *In* Warren, J. M., and Akert, K. (eds.): *The Frontal Granular Cortex and Behavior.* New York, McGraw-Hill Book Co., Inc., pp. 410-444.

Vaughan, E., and Fisher, A. E. (1962). Male sexual behavior induced by intracranial electrical stimulation. *Science* 137:758-760.

Velasco, M., and Lindsley, D. B. (1965). Role of orbital cortex in regulation of thalamocortical electrical activity. *Science* 149:1375-1377.

White, J. C. (1940). Autonomic discharge from stimulation of the hypothalamus in man. *Res. Publ. Ass. Res. Nerv. Ment. Dis.* 20:854-863.

Wolf, S. (1969). Central autonomic influences on cardiac rate and rhythm. *Mod. Conc. Cardiovasc. Dis.* 29:599.

CHAPTER 3

STEREOTAXIC SURGERY
AND HISTOLOGICAL
EXAMINATION OF THE BRAIN

A complete experiment on the brain of an animal involves both the chosen experimental surgery and the examination of the animal's brain in order to verify the exact anatomical regions affected. Neurophysiological studies can be made in the brain of an anesthetized, acute animal preparation or an awake, chronic preparation. The latter is preferred because it will have recovered from the effects of the anesthetic on its central nervous system and will be in a more normal, healthy state when the experiment is performed. An experiment upon a precise anatomical region of the brain is effected by mounting the animal in a mechanical device called a stereotaxic instrument, and implantation coordinates are chosen from a stereotaxic atlas of the brain so that the various implant devices can be inserted accurately into the chosen region.

The following figures illustrate the individual steps in the implantation of an electrode in the brain of a chronic animal preparation and describe how to examine the brain histologically after experimentation. Quite often a series of animals will have to be operated in order to achieve accuracy in implanting small neural structures. It is preferable to operate a single animal and examine its brain before operating the next one so that gross errors in stereotaxic placement are detected and eliminated. A stereotaxic atlas of the rat brain appears at the end of the book for use during the implantation procedure.

STEREOTAXIC SURGERY

Figure 3–1. ANESTHETIC DOSAGE

The most critical step in the surgical procedure is anesthetizing the animal. An animal is seldom lost as a result of brain trauma or infection; if a rat dies from the surgical procedure, it is more likely because it received a lethal overdose of the anesthetic. Any animal receiving a barbiturate anesthetic must first be accurately weighed to determine the dosage of the injection. Tables 3-1 and 3-2 show the dosages for the barbiturate anesthetic, sodium pentobarbital (Nembutal). For rats the usual dosage of this drug is 35 to 45 mg/kg body weight, but this dosage varies considerably among the different strains. Several rats of the strain being used should be injected with various dosages between 35 and 45 mg/kg to determine the one most effective.

Figure 3–1

TABLE 3–1   SODIUM PENTOBARBITAL DOSAGES WITH DRUG
CONCENTRATION AT 50 mg/cc*

Body Weight in grams	Dosage in cc	Body Weight in grams	Dosage in cc
200	0.168	350	0.294
205	172	355	298
210	176	360	302
215	181	365	307
220	185	370	311
225	189	375	315
230	193	380	319
235	197	385	320†
240	202	390	320†
245	206	395	320†
250	210	400	320
255	214	405	324
260	218	410	328
265	223	415	332
270	227	420	336
275	231	425	340
280	235	430	344
285	239	435	348
290	244	440	352
295	248	445	356
300	252	450	360
305	256	455	364
310	260	460	368
315	265	465	372
320	269	470	376
325	273	475	380
330	277	480	384
335	281	485	388
340	286	490	392
345	0.290	495	0.396

*The dosage rate of the drug per body weight is 42 mg/kg for rats under 400 gm and
40 mg/kg for rats over 400 gm.
†Transition period of change in dosage rate of the drug.

TABLE 3–2 SODIUM PENTOBARBITAL DOSAGES WITH DRUG
CONCENTRATION AT 64.8 mg/cc (1 grain/cc)*

Body Weight in grams	Dosage in cc	Body Weight in grams	Dosage in cc
200	0.130	350	0.226
205	133	355	230
210	136	360	233
215	140	365	236
220	143	370	240
225	146	375	244
230	149	380	246†
235	153	385	246†
240	156	390	246†
245	159	395	246†
250	162	400	246†
255	165	405	250
260	169	410	253
265	172	415	256
270	175	420	259
275	178	425	261
280	181	430	265
285	185	435	218
290	188	440	271
295	191	445	274
300	194	450	278
305	198	455	281
310	201	460	284
315	204	465	287
320	207	470	290
325	210	475	293
330	214	480	296
335	217	485	299
340	220	490	302
345	0.223	495	0.305

*The dosage rate of the drug per body weight is 42 mg/kg for rats under 400 gm and 40 mg/kg for rats over 400 gm.
†Transition period of change in dosage rate of the drug.

Figure 3–2. INTRAPERITONEAL INJECTION

Anesthetic drugs are injected into the peritoneal cavity (IP) for rapid action. In larger animals intravenous injection (IV) is preferred, but in rats such injections are difficult because the veins are so small and delicate. Notice in this figure that the surgeon's thumb and forefinger are touching each other underneath the chin of the rat. This method of holding the animal will prevent a severe bite. Notice also that the animal is firmly but gently supported by the other fingers wrapped around its chest. This will minimize the animal's squirming while the IP injection is being administered. The needle is jabbed sharply (not pushed) through the abdominal muscles. When properly done, the jab will not cause the animal to flinch. After the needle is jabbed sharply into the peritoneal cavity, withdraw the plunger slightly and look for the presence of blood to ensure that the needle tip is in the cavity rather than in some organ. Then inject the drug rapidly, and quickly remove the needle. The entire procedure of injection should be performed as soon as the animal is picked up; otherwise, it will squirm and fight back.

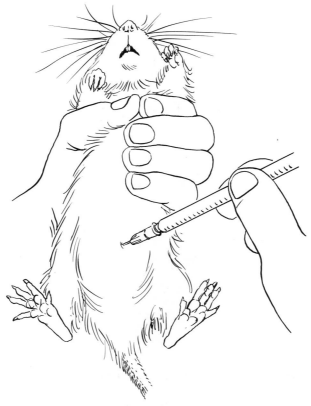

Figure 3–2

Figure 3–3. TAIL-PINCH TEST

Since precaution is taken to prevent lethal overdoses of anesthetics, some injections may result in slight underdoses, and the animal will not lose consciousness. A good test of the level of anesthesia is to pinch the tip of the animal's tail sharply with the thumbnail. If any leg flexion results, the animal is still conscious. If after 15 minutes or so the animal is still slightly conscious, administer additional anesthetic (about 15 per cent of the first dose). However, supplemention with barbiturate anesthetics is dangerous. A better method is to administer a supplemental dose of a gaseous anesthetic until complete unconsciousness occurs. This procedure is outlined next.

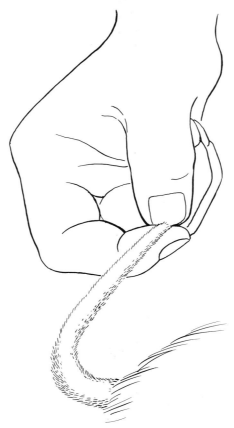

Figure 3–3

Figure 3–4. SUPPLEMENTAL ETHER ANESTHETIC

When an animal receives a slight underdose of barbiturate anesthetic, rather than giving more of the drug at the risk of a lethal overdose, administer a supplemental dose of ether. Ether is more accurately controllable than are barbiturate anesthetics, but has the disadvantage of requiring constant attention in its administration. It is advisable that the surgeon have an assistant to administer the ether, because it is most difficult to attend to the anesthesia and perform surgery simultaneously without endangering the animal. Pour the ether into a small jar (with a tight lid) with cotton on the bottom until the cotton is saturated. Force the head of the partially anesthetized animal into the jar; be careful not to let any liquid ether get into the animal's nostrils or come in contact with its skin. When the animal relaxes, lay it on its side with its head still partially in the jar, and begin the tail-pinch testing for complete anesthesia. Complete anesthesia usually occurs a few seconds after the animal relaxes. If the animal goes into convulsive breathing or stops breathing, remove it from the jar immediately and administer artificial respiration by squeezing quickly and firmly on the animal's chest at a rate of around two per second. After about 10 squeezes, see if the animal can breathe for itself. If it fails to breathe within five seconds, continue artificial respiration for another 10 squeezes, and so on. If the animal fails to respond, check its heartbeat by placing your thumb and forefinger on either side of its chest just below the armpits in order to feel the pulsations. The normal heart rate is a rapid thumping (around 6 beats/second). If the heart rate is slow and fading, the heart is dying and the animal will be dead in a few seconds unless oxygen gets into its circulation via artificial respiration. If everything goes well and the tail-pinch test shows the animal to be unconscious, quickly mount it in the stereotaxic ear bars, but continue to administer the supplemental anesthetic with a piece of ether-soaked cotton covering the nose. Be sure that all the liquid ether is squeezed out of the cotton. The cotton should be resoaked, wrung out, and readministered about every two or three minutes for maximum etherization. However, continual checking of the level of anesthesia by pinching the tail will indicate how often ether has to be readministered.

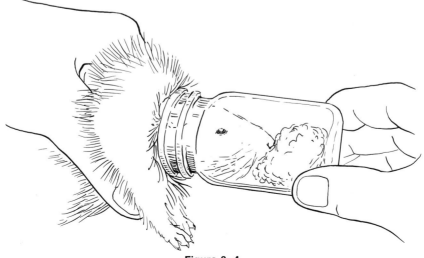

Figure 3–4

Figure 3–5. *SHAVING THE HEAD*

In order to minimize the possibility of infection and to facilitate surgery, shave the top of the animal's head from between the eyes to well behind the ears. Be careful not to get any hair in the animal's eyes. If you do, wipe it out with a wet cotton swab. Clean the clipped hair off the animal before putting it into the stereotaxic frame. A piece of hair in an open incision will be a source of irritation and infection during the healing process. Use electric barber's clippers for shaving the head.

Figure 3–5

Figure 3–6. MOUNTING THE ANIMAL IN THE STEREOTAXIC FRAME, 1

   With the animal's head pointing toward your left, grasp the tip of the animal's right ear with your right thumb and forefinger and grasp the animal's head firmly with your left thumb and forefinger. The fixed tip of the right stereotaxic ear bar is then directed into the auditory canal of the animal's right ear by moving the head sideways against the ear-bar tip. Bend over the animal in order to sight down into the ear canal. Once the ear-bar tip is in the canal, release the ear tip and firmly grasp the sides of the animal's neck while your left hand continues to support firmly the front of the head. If the right ear-bar tip is still in the ear canal, move the animal's head down and forward while maintaining slight pressure against the ear bar to assure that the tip stays in the canal. You should feel the tip slip suddenly into the auditory meatus and should not be able to move the animal's head up, down, forward or backward while maintaining pressure of the head against the ear-bar tip. Once this positioning is achieved, move on to step 2. Disregard any bleeding from the ear, for the ruptured ear drum or cut tissue will heal later, and the pressure of the ear bars will help to stop the bleeding.

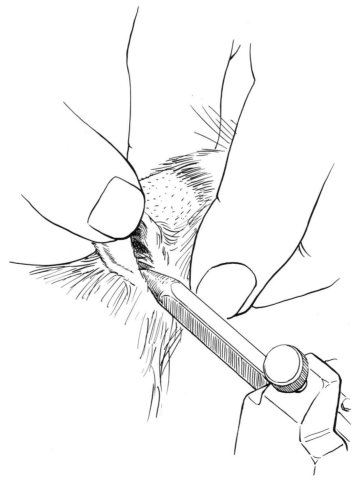

Figure 3–6

Figure 3–7.   *MOUNTING THE ANIMAL IN THE STEREOTAXIC FRAME, 2*

Mounting the animal in the stereotaxic ear bars is the most difficult manipulation in the entire surgical procedure and the most important step in determining the accuracy of placement of the implanted electrodes. Extreme care must be taken, therefore, to assure that each ear-bar tip is properly positioned in the auditory meatus. After assurance that the ear-bar tip is in the auditory meatus of the animal's right ear, release the animal's neck with your right hand while your left thumb and forefinger continue to support the head and apply pressure of the head against the ear-bar tip. Sight into the animal's left ear along the axis of the left ear bar. With your right hand, push the loosened left ear bar into the auditory canal and apply slight inward pressure. Move the left side of the animal's head down and forward and all around until you can feel the left ear-bar tip slip into the auditory meatus. With your right hand, continue to apply pressure on the left ear bar. Release the animal's head and, with your now free left hand, tighten the set screw of the left ear bar.

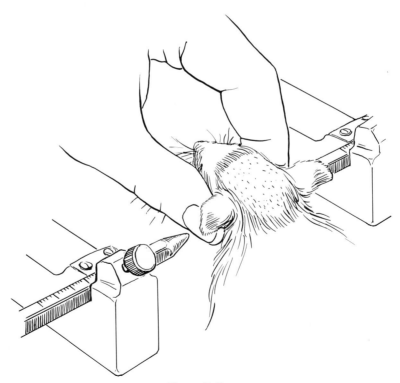

Figure 3–7

Figure 3–8. **MOUNTING THE ANIMAL IN THE STEREOTAXIC FRAME, 3**

To check whether the ear-bar tips are properly placed in the auditory meatuses, grasp the animal's nose and wiggle it back and forth firmly. If the animal is mounted properly, it will be impossible to wiggle it out of the ear bars, and the head will feel rigidly mounted.

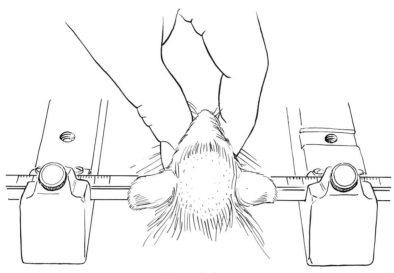

Figure 3–8

Figure 3–9. *MOUNTING THE ANIMAL IN THE STEREOTAXIC FRAME, 4*

If the right ear bar was not tightened in exactly the right place before the animal was mounted, the animal will not be centered between the ear-bar posts. To center the animal, gently press on both ear bars with the palms of the hands and loosen both of the set screws with the thumb and forefingers. Move the animal back and forth until it is centered. Retighten the set screws while pushing gently but firmly on the ear bars. Give a final nose wiggle and visual check to assure that the animal is properly mounted. Notice that the top of the ear bars is only slightly lower than the top of the head. Now mount the animal's teeth on the incisor bar and tighten the nose clamp. The top of the incisor bar must be exactly 5 mm above the center of the tip of the ear bars.

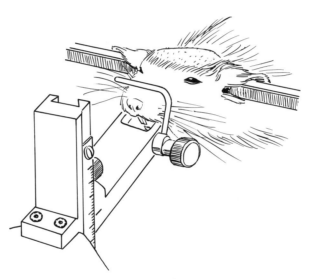

Figure 3–9

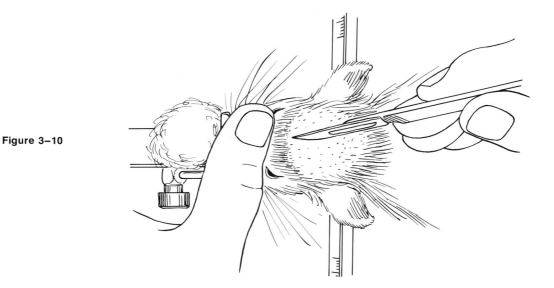

Figure 3-10

Figure 3-10. SCALP INCISION

Start the incision slightly behind the eyes on the midline. The incision is better controlled if the scalpel is pressed down through the skin and rests on the top of the skull bone and the skin is then drawn forward with the thumb.

Figure 3-11. LENGTH OF INCISION

The incision should be about 3/4 of an inch long, permitting both the bone suture junctions, bregma and lambda, to be seen in the wound. If the incision is too long, stitches will be needed to close the wound after the electrode connector is in place. Putting these stitches in is risky because the connector may be knocked loose.

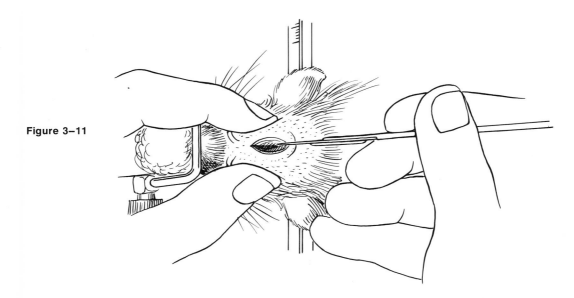

Figure 3-11

Figure 3–12. **SCRAPING THE PERIOSTEUM TO EXPOSE THE CRANIAL SUTURES**

After the incision is made, scrape the periosteal connective tissue which adheres to the bone. Scraping with a blunt edge such as the back of the scalpel blade will minimize bleeding because the capillaries will be crushed rather than cut smoothly. Continue to scrape and to remove the blood with a sterile swab until the cranial sutures are exposed and bleeding has stopped. Rub bone wax over the surface of the bone to stop any bone bleeding.

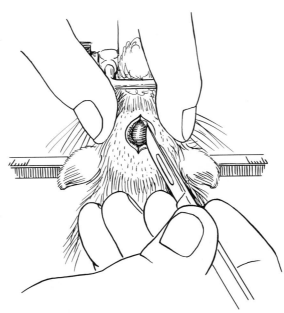

Figure 3–12

Figure 3–13. CRANIAL SUTURES: BREGMA AND LAMBDA

Drying the bone will make the cranial sutures more clearly visible. The point bregma (B) on the top of the skull is the intersection of the frontal and parietal skull plates at the midline. The point lambda (L) is the intersection of the parietal and interparietal skull plates at the midline. These points, as well as the auditory meatus, can be used for stereotaxic reference points. In the rats which were used to make the stereotaxic atlas in this book, the variability of the bregma point was less than that of the auditory meatus with respect to certain brain structures. Therefore, bregma is a better stereotaxic reference point. The distance between bregma and lambda can be used to form a correction factor for the stereotaxic coordinates for use on animals whose brains are considerably smaller than those of the standard size rats used to make the atlas (the standard rats used in the atlas weighted 350 to 500 gms and were 120 to 150 days old):

$$\frac{\text{Bregma} - \text{Lambda (your rat)}}{\text{Bregma} - \text{Lambda (atlas rat)}} \times \text{atlas coordinate} = \text{corrected coordinate}$$

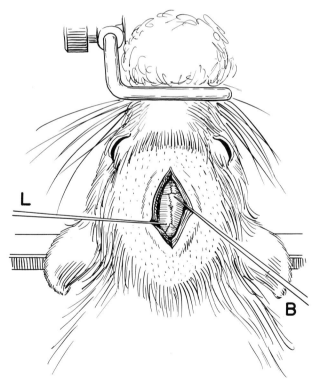

Figure 3–13

Figure 3–14. *MARKING AND CHECKING THE COORDINATE POINTS*
ON THE SKULL

Place a needle in the stereotaxic instrument and move it to the desired anteroposterior and lateral coordinates determined from the stereotaxic atlas. Carefully make a mark with a sharp needle on the top of the skull where the needle in the stereotaxic instrument indicates the trephine holes are to be drilled. Stereotaxic instruments are quite complicated, and mistakes are often made in determining the exact spot where the electrode is to be lowered through the skull into the brain. The positions of the needle marks which have been made on the top of the skull can be verified by measuring with a millimeter ruler the distance of the marks from bregma in the anteroposterior direction and from the midline in the lateral direction. Each of the stereotaxic atlas plates shows the anteroposterior distance from bregma as well as that from the auditory meatus, so verification is relatively easy to perform even though the latter coordinate system is used.

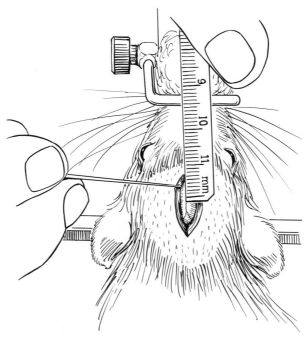

Figure 3–14

Figure 3–15. **DRILLING THE TREPHINE HOLES**

A hand-held pen vise with a small machine drill will make a smooth-walled trephine hole in the rat's skull. Furthermore, it has a built-in safety feature in that the drill can go through the skull only by the amount that the drill protrudes from the pen vise, thus eliminating the possibility of piercing the brain with the drill. Position the drill tip into one of the needle marks on top of the skull and manually twist it while applying a slight downward pressure. If the drilling is too fast, or if the drill tip has been dulled by contact with other instruments, the thin layer of bone cells on the underside of the skull bone will not be cut smoothly and may later deflect the electrode while it is being lowered through the hole. Allow the drill to extend out of the pen vise just slightly more than the thickness of the skull. Depending upon the age, type and sex of the rats being used, the extension of the drill out of the vise is from 1 to 2 mm. Drill the holes for the skull screws which anchor the electrodes to the head with a No. 56 machine drill, the correct size for the 0-80 stainless steel screws.

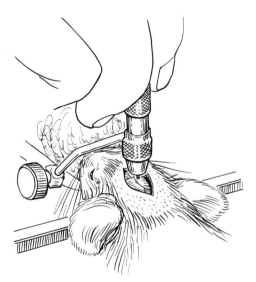

Figure 3–15

Figure 3–16. PLACEMENT OF THE SCREW HOLES

Drill trephine holes for each electrode that is to be implanted and for at least two skull screws. Drill the trephine holes for the skull screws as far lateral as possible and within a couple of millimeters of the electrode trephine holes. Be careful not to drill the skull screw holes too close to the electrode holes because the skull screw has a head on it which may cover the electrode hole. Also, the screw protrudes above the skull and may not allow the electrode carrier on the stereotaxic to be lowered far enough to reach the desired vertical coordinate. NEVER drill closer than 0.5 mm to the midline longitudinal suture. Directly below this suture is the sagittal sinus, piercing of which may result in the animal's bleeding to death or in brain damage as a result of vascular hemorrhage. If you want to implant an electrode tip directly under the sinus vein, angle the direction of insertion of the electrode.

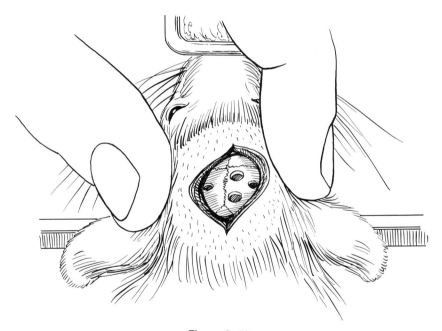

Figure 3–16

Figure 3–17. SLICING THE DURA MATER

The dura is a very tough membrane surrounding the brain and must be cut before the electrode is lowered into the brain; otherwise, the dura will bend and deflect the electrode tip. A very sharp hypodermic needle is ideal for cutting the dura. Do not jab the needle downward through the dura; rather, slice it horizontally as illustrated in the insertion in the figure. If jabbed downward, the needle will only depress the dura into the brain instead of piercing it. If a sharp needle is inserted properly, you can feel the dura being sliced when the needle is moved back and forth from one side of the trephine hole to the other. Be extremely careful not to pierce the sinus vein while performing this step. If the sinus vein is inadvertently pierced, the bleeding is most easily stopped by filling the trephine hole with a piece of cotton. The bleeding may begin again when the cotton is removed to lower the electrode; if it does, quickly fill the trephine hole this time with bone wax. When the bleeding stops, the electrode then can be lowered through the soft bone wax. This step in the operation of piercing the dura also allows a check of how well the trephine holes have been drilled. If you can feel a jagged edge at the bottom of the hole, the thin lower layer of bone cells has not been drilled through. Lengthen the drill protrusion from the vise another 0.5 mm, drill again, and see if the jagged edge is eliminated. If not, it can sometimes be cut smooth with the sharp dura needle, but this procedure dulls the cutting edge, and a new needle will be needed for further dura slicing. Do not cut the dura beneath a skull screw trephine hole.

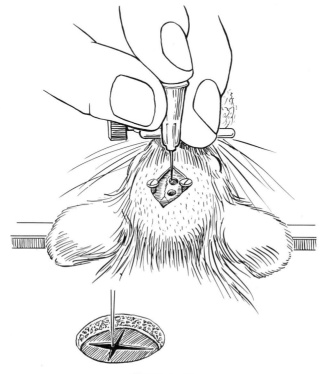

Figure 3–17

Figure 3–18   *ATTACHING THE SKULL SCREWS*

Always use stainless steel screws. Corrosive metals will damage the meninges and cortex. Small jeweler's screws with as many threads per inch as possible are desirable; the 0-80 × ⅛ stainless steel machine screw with a fillister head is best for use in the rat. To insert the screw, hold it with fine forceps and slide the end into the trephine hole, making sure the screw is perpendicular to the top of the hole. Then, without moving the position of the screw, put the tip of the screwdriver into the screw slot, apply a very slight pressure, and slowly twist the screw until it begins to self-tap its threads into the bone. Once the screw has started, it has advanced about 0.5 mm into the skull, which leaves only a turn or two more to be threaded. In an 0-80 screw there are about three turns per millimeter. Do not twist the screw in much farther than the thickness of the skull. If your hands are shaking, rest them on a firm support while inserting the screw.

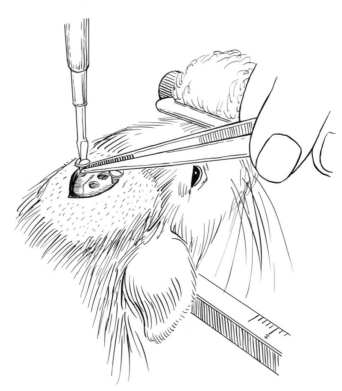

Figure 3–18

Figure 3–19. *ALIGNMENT OF THE ELECTRODE*

    After the electrode has been inserted in the electrode holder of the stereotaxic instrument, it must be perfectly aligned with the axis of the dorsal-ventral probe drive. Align the attached electrode with a post located on the stereotaxic frame which has already been placed in perfect alignment. Bend the electrode into alignment with the post by inspecting and aligning it in two directions 90 degrees apart. Next, place the tip of the electrode on the reference point of the stereotaxic instrument and record the reference coordinates for computation of the implant coordinates.

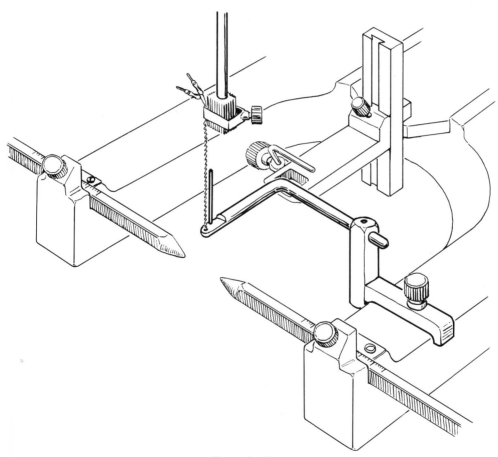

Figure 3–19

Figure 3–20. **IMPLANTING THE ELECTRODE**

Align each electrode in the dorsal-ventral axis after it is mounted in the electrode carrier. If no zero aligning device is available, alignment can be made by visually aligning the electrodes with various perpendicular parts of the stereotaxic frame. Before the electrode is mounted in the holder, be sure it is straight. Sight down the electrode and bend it straight with your fingers. Sterilize the electrode by soaking it in 70 per cent alcohol for as long as possible before it is implanted. Be sure the alcohol has evaporated off the electrode before you implant it. Lower the electrode through the trephine hole in the skull and through the sliced dura mater into the soft brain tissue to the desired coordinate point. A single electrode of 0.015-inch diameter or less does not do any appreciable damage when implanted in most parts of the brain. However, it is possible to damage small respiratory or metabolic centers in the hypothalamus with such an electrode; death may ensue, but rarely does. Electrodes lowered through the septal region of the brain may produce small lesions in this area, resulting in a ferocious "septal animal." Be very careful when first handling an unfamiliar animal because it may have such a lesion. It is best to implant only two electrodes in a single rat brain. Four electrodes is the maximum number that should ever be implanted in a single rat brain because additional ones may produce enough damage and trauma to cause bursts of seizure activity in the brain.

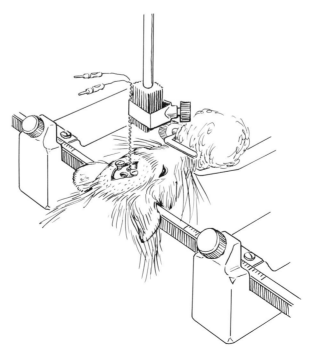

Figure 3–20

Figure 3–21. **REMOVING THE ELECTRODE FROM THE STEREOTAXIC INSTRUMENT**

There are two methods of holding the implanted electrode in place before it is removed from the stereotaxic instrument: (1) In the *fastest* method, the implanted part of the electrode is held in place firmly with forceps while the remainder of the soft, untempered electrode wire is bent down to the skull, run over the surface to the skull screw, and wrapped around the screw several times for firm support. The electrode is grasped with fine-tipped forceps which rest in two places, one on the skull and the other on some support provided for this purpose. (2) In the *best* but slower method, the electrode is cemented in the trephine hole and to the skull screw before it is removed from the stereotaxic frame. This method eliminates the possibility of any movement of the electrode tip which might dislocate it or damage the brain tissue. Before the dental cement is applied, a small piece of Gelfoam, dipped in physiological saline and then pressed nearly dry, is placed around the electrode to keep the toxic dental cement off the brain.

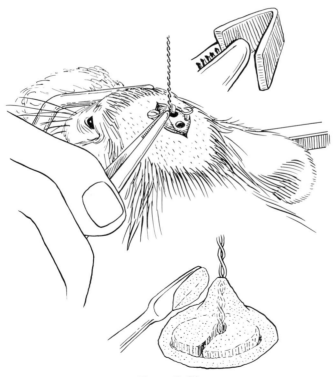

Figure 3–21

Figure 3–22. **THE FAST METHOD OF ANCHORING THE ELECTRODE WIRES TO THE SKULL**

To use the fast method, bend the dangling end of the implanted electrode over the tip of the forceps toward the screw to which it is to be attached. Wrap the wire twice around the screw shank with your fingers. Next, press the wire that is between the forceps and the screw firmly down to the surface of the skull with the blunt end of a small wooden stick. Carefully remove the forceps from underneath the bent part of the electrode. Implant the second electrode and wrap it around the other screw. If more than two electrodes are used, wrap two electrodes around the same screw. After all electrodes have been implanted, and Gelfoam has been placed in the trephine holes (see Fig. 3-21), apply the dental cement. Be careful not to build up any cement over the top of the electrode wires traveling from the trephine holes to the screws. The electrode connector must fit on top of these implanted electrodes. If contacts are not on the ends of the electrodes, crimp them on after the dental cement has completely hardened. It is best to attach the contacts before implantation in order to save time during the surgery, but first be sure to determine the precise length of the electrode.

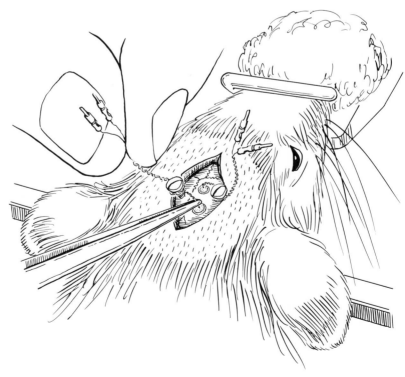

Figure 3–22

Figure 3–23. **INSERTING THE ELECTRODE CONNECTORS INTO THE CONNECTOR BASE**

After the electrodes are firmly cemented to the skull screws, the free connector ends can be moved without disturbing the electrode tip. With hemostatic forceps, insert the first half of each contact into one of the holes in the connector base, making sure to insert the contacts into the correct end of the base. In order to remember the postion of each electrode in the base, keep a written record. If implanting several animals with the same types of stimulating and recording electrodes, be consistent about the positions of the contacts in the connector base. Then you can use the same stimulating and recording cables for all the animals.

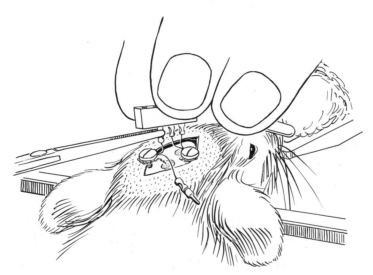

Figure 3–23

Figure 3–24. LOCKING THE CONTACTS IN THE CONNECTOR BASE

After the contacts are partially inserted into the connector base, force them all the way in with the forceps, as illustrated. They will lock permanently into the base when forced all the way in. Let the forceps tip just barely catch one edge of the small contact so that the electrode wire is not caught between the contact and the forceps tip, thus damaging the electrode insulation. If there is any excess wire, carefully fold it alongside the connector base or underneath it; but be careful not to cause the connector base to stick up too high on top of the animal's head. The higher the base, the greater the leverage force it will exert on the skull screws when the recovered animal bumps its head on the side of its cage. The bottom of the connector base should not be separated more than 3/32 inch from the top of the skull; otherwise, the animal may knock off the whole implant pedestal.

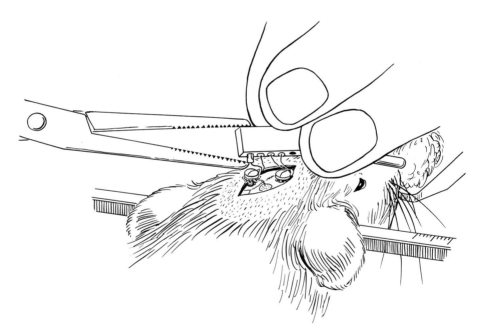

Figure 3–24

*Figure 3–25. **CEMENTING THE CONNECTOR BASE TO THE SKULL SCREWS, 1***

Dental cement can be applied in several ways. The dry powder can be squirted onto the top of the skull as shown, and then the liquid can be applied with a syringe and needle. This method is fast, but mixing the cement in a dish before applying it gives better control of the consistency of the cement.

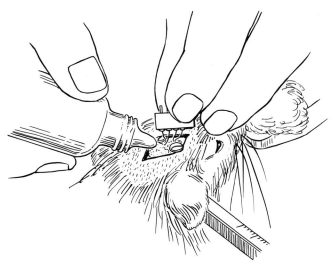

Figure 3–25

*Figure 3–26. **CEMENTING THE CONNECTOR BASE TO THE SKULL SCREWS, 2.***

Place enough of the cement underneath the connector base so that the base can be pressed down into the mushy cement. Spread the cement completely around both of the skull screws. Care should be taken to ensure that the bare metal of the contacts on the underneath part of the connector base is completely covered by the cement and that no bare spots on the electrode wires are touching together. If body fluids can ever seep between a stimulating contact and a recording one, very large stimulus artifacts will occur.

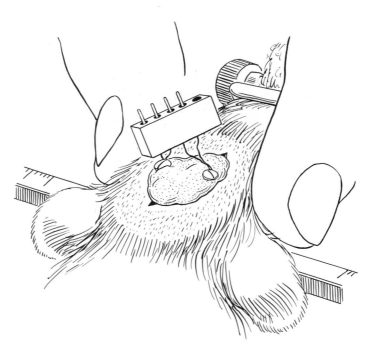

Figure 3–26

*Figure 3–27   CEMENTING THE CONNECTOR BASE TO THE SKULL
SCREWS, 3*

After the connector base has been held down in the cement long enough for it to harden into a rubbery consistency that is still soft enough to mold, pull the skin up over the cement. If the cement has hardened too much, so that there are sharp edges where the cement has run down to the cut edge of the skin, cut off these sharp edges immediately with scissors. Sharp edges will later irritate the skin and lead to infection.

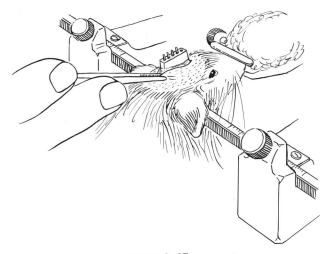

Figure 3–27

Figure 3–28. ***CEMENTING THE CONNECTOR BASE TO THE SKULL SCREWS, 4***

While the cement is still soft enough to mold, press the fingers down on the outside of the skin, thus molding the cement underneath it. Smooth out all the rough edges and bumps in the cement, and keep the edges of the wound pressed close to the connector base. If the cement has hardened too much, a few drops of the liquid will soften it a little. If too much cement has been applied, the skin will not pull up close to the connector base. Hold the connector base and skin in place for about three minutes until the cement has hardened considerably. If the incision was made too long, a single stitch may be put in the skin just posterior to the connector base. If a stitch is necessary, put it in after the cement has completely hardened.

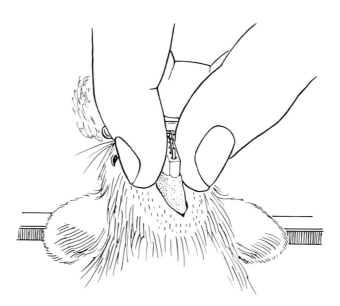

Figure 3–28

Figure 3–29. **POSTOPERATIVE RECOVERY**

After the cement has completely hardened, remove the animal from the stereotaxic frame. Apply antiseptic powder (e.g., sulfathiazole) or antibiotic ointment (e.g., miltrafurazone) copiously around the wound. If there is any bleeding from the ears, place some antibiotic well into the ear canal. Keep the animal warm because barbiturates prevent the animal from maintaining its body heat. House each chronically implanted animal in a separate cage; if housed together, they will begin within a matter of minutes to chew on each other's connector bases. It usually takes four to seven days for the animal to recover completely from the surgery.

Figure 3–29

Figure 3–30.   ANTIBIOTIC TREATMENT

Many authorities report that systemic antibiotic treatment of postoperative infection is unnecessary in the rat. The rat apparently has sufficient mechanisms to ward off any infection without outside aid. However, in phylogenetically higher animals, antibiotic treatment is usually required, and in primates sterile surgery is necessary. Most antibiotics are administered intramuscularly. The biceps femoris muscle is usually the muscle chosen because it is the largest. This muscle is held between the thumb and forefinger as illustrated, and the needle is jabbed sharply into the muscle. After insertion of the needle, the syringe plunger should be pulled out slightly to see if any blood is aspirated. Antibiotics should never be injected directly into a vein. If blood is aspirated, jab the needle in farther and reaspirate. The antibiotic should be injected and rubbed into the muscle with the thumb and forefinger. A dosage of 50 mg/kg body weight of chloromycetin given three times a day for four days is sufficient for the rat. Chloromycetin is the preferred antibiotic because it crosses the blood-brain barrier and gets directly into the brain to fight any infection caused by the electrodes. It is difficult to give an IM injection to a rat without assistance, so, to prevent any serious rat bites, it is recommended that the rat be either held by an assistant or wrapped in a towel.

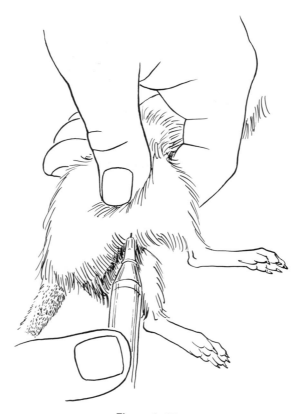

Figure 3–30

Figure 3–31. *ATTACHING THE ANIMAL'S CABLE TO ITS CONNECTOR*

While the animal is recovering from surgery, handle it at least once a day so that it will become docile and easy to handle later during the experiments. Note in this illustration how to hold the animal while attaching the cable connector. Hold its head between your index and middle fingers and slip your thumb and ring finger behind the forelegs of the animal. If the animal is held firmly in this manner it cannot move its head or bite your fingers. The flexible recording and stimulating wires of the animal's cable are first cemented to the connector with dental cement, and then silicone rubber or tape is applied so that the wires do not break off at the point they emerge from the dental cement. The smallest flexible wire available should be used in the cable, and the diameter should not exceed 28 gauge. Stranded copper wire with as many single strands as possible, covered with a flexible polyvinyl insulation, is preferable. Recording artifacts will occur when the animal moves rapidly or shakes its head if the contacts of the connectors are able to move in the slightest. For this reason it is best to use masking tape to secure the connectors in place when recording through the cable. For stimulating or lesioning, the connectors can be secured by a small machine screw which is easier to attach and detach. Rats will chew on the cable wires if left alone and unoccupied in the experimental chamber. A small, lightweight spring slipped over the part of the cable close to the animal will prevent the animal's damaging it.

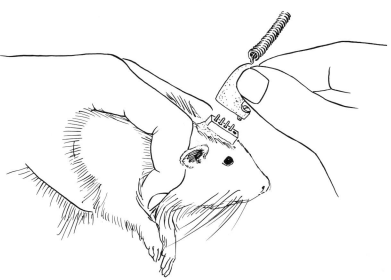

Figure 3–31

HISTOLOGICAL EXAMINATION OF THE BRAIN

Figure 3–32. CARDIAC PERFUSION

Before brain tissue can be cut into thin sections and subjected to histological examination, it must first be fixed and hardened. In small animals, such as the rat, it is not always necessary to perfuse the brain by injecting the fixative into the animal's vascular system because the fixative can diffuse from the outside of the exposed brain into the center in a few days and completely harden the tissue. However, this method causes the brain to shrink by as much as 30 per cent (using the usual 4 per cent formaldehyde fixative solution), and distorts the shape of the brain. Furthermore, because of the shrinkage, this method requires that the electrodes be removed before the tissue is fully fixed, thereby permitting the unfixed tissue to close around the electrode tracks, making them smaller and more difficult to locate during sectioning. The perfusion method has many advantages over this more expedient soaking technique: (1) more rapid fixation (the brain can be sectioned immediately after perfusion); (2) minimum distortion and shrinkage of the tissue (only 6 per cent using a 40 per cent solution of formaldehyde fixative); (3) better fixation and localization of the electrode tracks (the exact outline of the electrode can be observed in the tissue sections).

In larger animals, such as the cat, the fixative can be injected through a cannula directly into the easily accessible common carotid artery, thus forcing the blood out of the brain and back into the venous system of the lower body. However, this technique is difficult to perform in the rat because its arteries are too small to cannulate easily. Injection of the fixative into the brain through the heart is easy to perform, though it is somewhat messy. This technique is as follows:

1. Anesthetize the animal with a lethal overdose of a barbiturate drug.

2. As soon as the animal is surgically anesthetized, make an incision through the skin along the left side of the animal's sternum.

3. Peel back the skin and then push the pointed lower blade of a pair of scissors underneath the lower boundary of the rib cage and cut along the sternum through the soft ribs and muscles.

4. Make a lateral incision along the lower boundary of the rib cage and turn a flap, exposing the heart. Cut off this flap. The procedure must be carried out very rapidly because as soon as the thoracic cavity is punctured, the animal is unable to respirate. The brain should be perfused with fixative before cyanosis occurs.

5. As soon as the beating heart is exposed, grab the lower tip of the animal's right ventricle with mouse-tooth-tipped forceps and make a large incision in it to drain the incoming venous blood.

6. Quickly inject around 10 to 20 ml of physiological saline into the animal's left ventricle, as shown, to flush out most of the blood in the

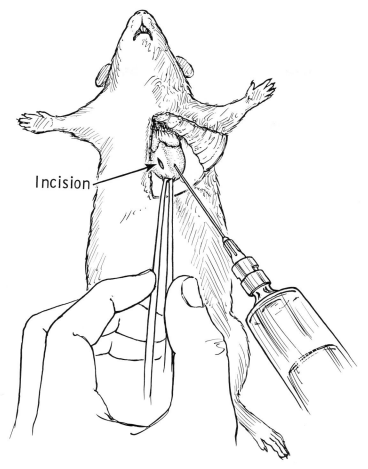

Incision

Figure 3–32

heart and brain. Insert the large hypodermic needle (20 gauge) in the lower part of the left ventricle, as shown, and be very careful to prevent the needle tip's puncturing the septum separating the ventricles. *Keep your eyes fixed on the heart* to prevent inadvertent movement of the needle tip.

7. Remove the needle and then quickly insert another needle *through the same insertion hole* and inject around 40 to 50 cc of 40 per cent formaldehyde (undiluted). The lower jaw of the animal should move immediately if the fixative is entering the brain.

Figure 3–33.   *BLUE-DOT STAINING TECHNIQUE*

It is sometimes difficult to locate the tip of an electrode track in the brain tissue during sectioning. A blue stain can be made to appear in the tissue around the electrode tip if stainless steel electrodes are used, and this blue dot can be looked for during the cutting. Passing direct current through the electrode tips into the brain will drive iron ions from the metal into the tissue. The tissue will turn a dark blue in the presence of potassium ferrocyanide (potassium ferricyanide also works, but is less effective). Inject the potassium ferrocyanide into the brain tissue by dissolving it in the fixative before perfusion. Make a saturated solution of the potassium ferrocyanide in the 40 per cent formaldehyde by putting in a few more crystals than will dissolve. Pass the current through the electrode after the animal's circulation has stopped, or preferably after the animal has been perfused. If the ions are driven into the tissue while the animal is alive and his circulation is intact, they will diffuse away from the tip, as shown in the right blue-dot stain in this figure. If the current is passed immediately after circulation has stopped, then the ions will be confined to the border of the electrode track as seen in the left blue-dot stain. Only a small amount of current passed for a short time is necessary to deposit the iron ions in the tissue (5 to 10 millicoulombs). Current can be passed for around 30 seconds from an ohmmeter with a 9-volt battery turned to the largest resistance scale. Attach the positive lead to the electrode and the negative lead to the animal's body to drive iron ions from the electrode tip.

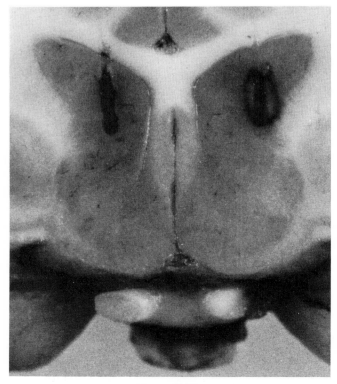

Figure 3–33

Figure 3–34. EXTRACTING THE BRAIN, 1

After the animal has been perfused, decapitate it with a pair of bone cutters or electrician's wire cutters. Cut the skin posterior to the connector base with a scalpel, and peel the skin away from it and forward. Cut off the peeled skin and distal nose bone. Insert one blade of the bone cutters in the mouth, and cut off the lower jaw. Be careful not to knock the connector base loose from the skull. Peel and cut off most of the attached muscle until you have only the brain case remaining, as shown.

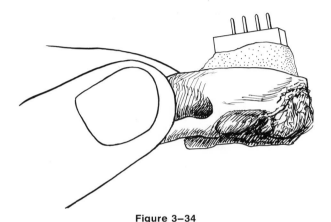

Figure 3–34

Figure 3–35. EXTRACTING THE BRAIN, 2

Next, begin to chip away the bone with a pair of rongeurs, starting at the foramen magnum where the spinal cord enters the brain case. Gently depress the underlying brain with one rounded tip of the rongeurs, ease in between the dura and the skull, and then cut a small chip. It is better to make numerous small cuts, as larger ones tend to rip the bone rather than cut it.

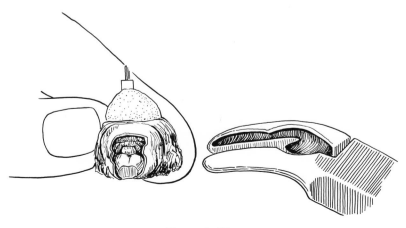

Figure 3–35

Figure 3–36. EXTRACTING THE BRAIN, 3

Chip away the bone covering the back of the cerebellum and the sides of the cerebrum, as shown. Note the thin dura mater covering the brain and try not to cut this during the chipping. If you are careful, however, you can cut this membrane without damaging the underlying tissue. After removing the back and sides of the brain case, cut off the bottom, as shown. Be very careful at this point if you want the pituitary gland intact. If the brain was not well perfused and is soft, it should be placed in 4 per cent formaldehyde for 12 to 24 hours. If it was well perfused, cut off any remaining dura with iris scissors, and then cut through the olfactory bulbs.

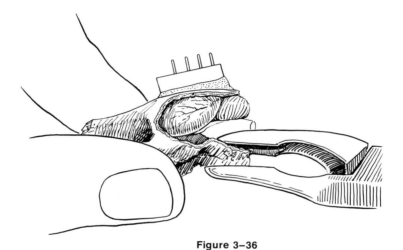

Figure 3–36

Figure 3–37. REMOVING THE ELECTRODES

The brain case should now be entirely chipped away, except for the front part of the olfactory sinus bone, containing the olfactory bulbs, and the top portion of the skull which is attached to the connector base. If the electrodes were all implanted perpendicular to the top of the brain case and parallel to each other, then the brain can be pulled straight down from the top, as shown. If the electrodes were implanted at angles to one another, they will be difficult to remove without damaging the tissue. However, the flexibility of the electrodes and the elasticity of the fixed tissue permit this same method of removal if the electrodes are not at too great an angle with one another.

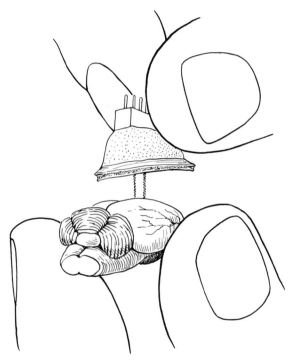

Figure 3–37

Figure 3–38. **BLOCKING THE BRAIN**

Only that part of the brain which contains the lesion or electrode tracks needs to be subjected to histological examination. In order to cut out that particular block of tissue without cutting into the areas of interest inside it, cut the anterior and posterior surfaces of the block parallel to the electrode tracks in the vertical stereotaxic plane. If the brain is allowed to rest upside down so that the dorsal parts of the cerebrum and cerebellum contact a flat horizontal surface, then the vertical stereotaxic plane will be at a 22 degree angle posterior to the perpendicular of the horizontal resting surface (see the illustration). Note in the figure that "A" is at the anterior part of the brain, and "P" at the posterior, and that the brain is resting upside down.

Using a cutting guide and razor blade, as illustrated, make the first cut in the very anterior portion of the brain, far away from any of the electrode tracks. Then look at the dorsal surface of the brain and note how far this cut is from the nearest electrode hole. Return the brain to its inverted position and make the thickness of the next block cut off less than this distance, leaving 2 to 3 mm of tissue anterior to the most anterior electrode track. Position the next cut so that it will be 2 to 3 mm posterior to the most posterior electrode track in the block. Make each cut carefully so that the block is bilaterally symmetrical. If the brain is carefully blocked, the thin sections to be cut from it will look exactly like those in the stereotaxic atlas, and you will be able to identify the neural structures easily.

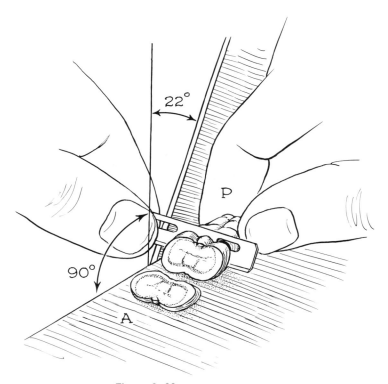

Figure 3–38

Figure 3–39. **PEDESTAL FREEZING MICROTOME**

There are several types of freezing microtomes on the market, and the principles of operation described here for the pedestal version are also valid for the other types. A ground-glass pedestal microtome with a hand-held knife is used in conjuction with a freezing unit that is simple to construct. The freezing unit consists of a lead freezing plate cemented with epoxy to a piece of cryogenic tubing which contains within it a washer, a spring and a spring lock. After the tissue is frozen to the lead plate, dry ice is inserted in the freezing unit and spring-loaded to keep the tissue frozen while it is being sectioned. The maximum thickness of the tissue block that can be sectioned is around 8 to 10 mm, and the maximum diameter is around 20 mm. This microtome is ideally suited for sectioning the whole rat brain and smaller blocks of other larger brains. It is inexpensive to purchase and withstands rough use. Because the knife edge is supported very close to the tissue block by the ground surface, very thin frozen sections can be cut.

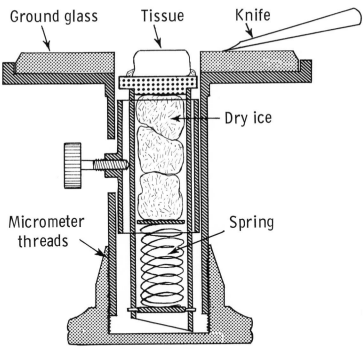

Figure 3–39

Figure 3–40.   FREEZING THE TISSUE BLOCK TO THE FREEZING UNIT, 1

Insert dry ice in the freezing unit and press it against the inside part of the freezing plate with a rod, or with your finger and the washer, until frost starts to form on the outside part of the freezing plate. Then apply a small amount of a 10 per cent solution of ethanol and make a liquid slush. Regulate the degree of cooling by controlling the force exerted on the dry ice to press it against the freezing plate. Do not let the slush run down the sides of the freezing plate.

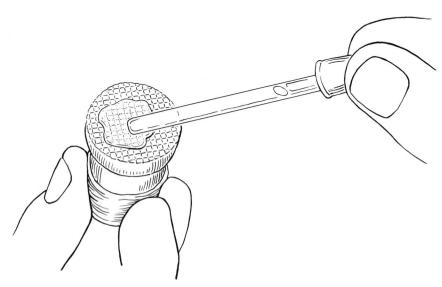

Figure 3—40

*Figure 3–41. **FREEZING THE TISSUE BLOCK TO THE FREEZING UNIT, 2***

Press the tissue block down lightly into the slush until it rests flat on the surface of the freezing plate. Next, exert a hard pressure to force the dry ice against the freezing plate, and allow the slush and tissue block to freeze. For additional support, slowly apply 10 per cent ethanol around the sides of the tissue block, as illustrated. A piece of dry ice can be placed on top of the tissue block to speed up the freezing process. After the tissue has frozen solid and turned to a lighter color, refill the freezing unit with dry ice, insert the spring and washer, and then lock them in place. In using microtomes which have fixed freezing plates and knife holders, the tissue block must be properly oriented on the freezing plate so that the knife cuts through the tissue in the desired direction. It is preferable to cut *into* an area of interest on the block rather than *out of* it because often, as the knife blade leaves the frozen tissue, it does not cut it sharply and may tend to tear the tissue at this border. It is better to cut out of the ice surrounding the tissue than out of the tissue itself.

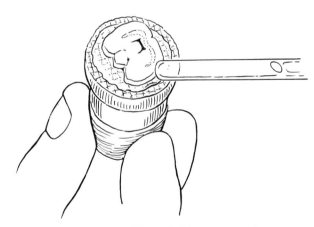

Figure 3–41

Figure 3–42.   CUTTING THE SECTIONS

After the tissue block has been frozen to the freezing plate, the freezing unit refilled with dry ice and the spring locked into place, insert the freezing unit in the microtome and lock it into place. The freezing plate is eccentric with respect to the dry ice tube, and the part that hangs over the side of the tube should be pointed toward the set screw before the freezing unit is locked into place. Lower the microtome drive until the top of the tissue block is even with the ground-glass surface. Cut the first pieces of the uneven top of the tissue block with the knife. Press the knife down firmly on the ground glass, holding it at a fixed angle, and slide it across the surface of the glass with a slight crossward slicing movement. Lubricate the surface of the glass with water. If the first small pieces of tissue flake and crumble when cut, then the tissue is too cold and hard. When cut, the tissue on the top of the block should just barely be frozen. A few drops of 95 per cent ethanol will melt the top surface slightly and keep the sections from flaking.

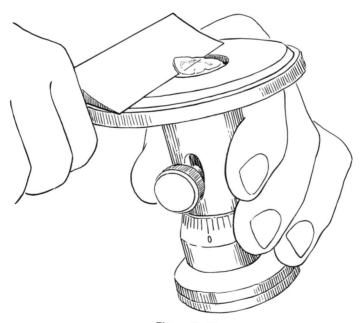

Figure 3–42

Figure 3–43. HOLDING THE KNIFE

The knife must be held at a fairly constant angle of about 30 degrees to the ground-glass surface, as illustrated. If the knife edge is *straight* it will always stay in the same plane of the ground-glass surface, no matter what the angle is. The vector force exerted on the knife parallel to the horizontal plane of the ground glass is the tissue-cutting force, and the downward vector force keeps the knife edge in contact with the glass. If the knife angle is less than 30 degrees, there is a tendency for the lessened downward vector force to be insufficient to counter all inadvertent upward muscular jerks of the arm made while cutting. If the knife edge is not perfectly straight, then any change in the knife angle will cause a change in the plane of the cutting edge and will result in sections of varying thickness. The hollow ground side of the knife edge should always be upward, and the plane side downward, to prevent irregularities in the ground side from changing the cutting plane. Check your knife edge for straightness and bevel before cutting. Always return the knife to its holder when it is not in use, and never lay it down on the table top. Wash the knife with 95 per cent ethanol after use, and wipe a thin coat of oil over both sides of the blade and the edge. Send dull knife blades to professional sharpeners who know the intricacies of microtome knife sharpening.

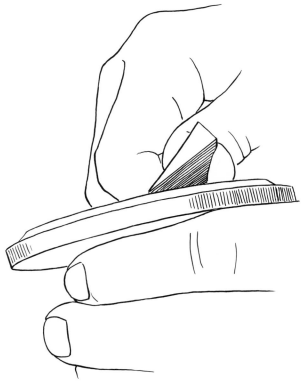

Figure 3–43

Figure 3–44. FLOATING THE CUT SECTION IN WATER

One of the advantages of the hand-held microtome knife is that the entire blade can be dipped in water to float off the cut section, as illustrated in this figure. Microtomes which have fixed knife blades must have the section wiped off with a brush or the finger and then floated off in water. If the section floats off in many pieces it may be that the tissue block is too cold and the section is cracking while it is being cut; or the lesion, electrode track or ablation may have cut the interconnecting fibers in the section which would hold the pieces together. In this latter case the tissue can be embedded in gelatin (see Table 3-3, Gelatin Embedding), which will hold all the pieces together.

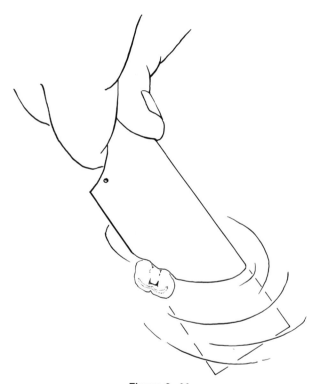

Figure 3—44

TABLE 3–3   GELATIN EMBEDDING

   This is the procedure for embedding tissue to be sectioned with a freezing microtome.

   The gelatin-embedding technique will enable one to section pieces of tissue that are not interconnected in the plane of cutting and would separate when floated in water during the mounting process. After the tissue has been fixed for 24 hours, or has been well perfused, wash it in water a few minutes, and then cut into the blocks which will be sectioned. The blocks are embedded in gelatin as follows:

1. Soak in 2½% gelatin solution at 37° C	4 to 24 hours
2. Soak in 5% gelatin solution at 37° C	4 to 24 hours
3. Soak in 10% gelatin solution at 37° C	4 to 24 hours
4. Place tissue in liquid 10% gelatin solution at 37° C and harden in a refrigerator	1 to 2 hours
5. Cut hardened gelatin blocks with tissue inside and fix them in 10% formalin	4 to 24 hours
6. Freeze, section, mount and stain like regular frozen sections	

   The 37° C temperature is not critical; what is desired is to keep the gelatin melted in a liquid state. A water-filled electric skillet turned as low as possible will keep the gelatin solutions melted nicely. The pure gelatin sold in grocery stores will work well in this method. Do not forget to put fixed gelatin blocks in 10% ethanol before sectioning.

Preparation of Solutions:

   2½% gelatin solution:
pure gelatin	2.5	gm
water (distilled)	100	ml

   5% gelatin solution:
pure gelatin	5	gm
water (distilled)	100	ml

   10% gelatin solution:
pure gelatin	10	gm
water (distilled)	100	ml

Figure 3–45   MOUNTING THE SECTIONS ON MICROSCOPE SLIDES

Store each of the serial sections cut from the block in individual compartments of polyethylene ice cube trays filled with water. After all the sections are cut, mount the ones of interest on microscope slides for closer histological examination. Egg albumin must first be applied to the microscope slides if the sections are to be stained. A light coat of albumin wiped on the slide and allowed to dry will permanently cement the sections to the glass slides throughout the staining procedures. The sections can be moved from the storage compartments with a small brush and placed in a larger bowl for floating and mounting on the slides. To mount a section, submerge part of the slide in the water in the bowl containing the section, as illustrated. Gently move the floating section over to the slide with a brush, and hold it down against the slide while the slide and section are slowly withdrawn from the water. Do not let any part of the section hang over the edge of the slide while it is being withdrawn.

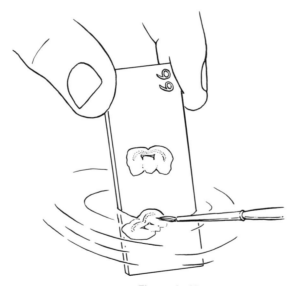

Figure 3–45

Figure 3--46. ADJUSTING THE SECTION ON THE SLIDE

After withdrawing the section and slide from the water, the section will be wrinkled and perhaps overlapped. Gently smooth out the section with a brush, as illustrated, and adjust its position on the slide. If an electrode track is in the section, it will pull apart, showing a wide gap. Adjust the section with the brush until the electrode track appears to be the same width it was when observed on the block during cutting. If the photographic histological technique is going to be used, put a few drops of glycerin on each section to keep it from drying out. If the sections are going to be stained, allow the sections to dry thoroughly before immersing them in the stain; otherwise, the albumin will not set and permanently adhere to the slide.

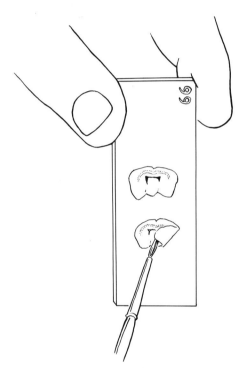

Figure 3--46

Figure 3–47. *PHOTOGRAPHIC HISTOLOGIC TECHNIQUE*

This technique is recommended for rapid histological verfication of electrode placements and the extent of lesions and ablations. The resulting photographs will show myelinated fibers in distinct detail but will not reveal individual nuclear groups. Each photograph in the stereotaxic atlas was made using this method, and the sections were later stained with a Nissl stain for identifying the subcortical nuclei. This photographic method (Guzman-Flores et al., 1958) uses the thin frozen sections (50 to 100 microns thick) in the same manner as a negative in a photographic enlarger is used to make a positive print. Use high contrast photographic paper for best results. For an enlargement of 10 times from a section 75 microns thick, expose Kodabromide F-5 paper at an iris setting of F4.5 for approximately 10 seconds. (Vary the exposure time until the best results are obtained.) Highly myelinated structures such as the brain stem require considerably longer exposure times. Be sure the section is wet before exposing it in the enlarger. If water is used to keep the slides wet, it must be reapplied periodically so that the sections will not dry out and shrink. If glycerin is used it will not dry out, but it must be removed before the sections are stained. It is recommended that the sections be stained with thionine or cresyl violet after the photographic prints are made so that the nuclear groups can be recognized in the sections. Staining shrinks and distorts the tissue, and the photographic prints should be relied upon for undistorted measurements in the tissue. See Tables 3-4 through 3-8 for the various staining techniques.

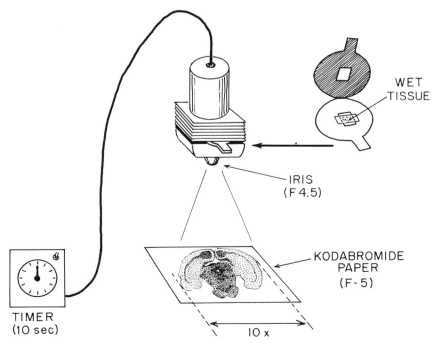

WET TISSUE

IRIS (F 4.5)

KODABROMIDE PAPER (F-5)

TIMER (10 sec)

10 x

Figure 3–47

TABLE 3–4   NUCLEUS STAIN

This buffered Nissl material stain for identifying nuclear structures in the brain and spinal cord is the routine cell body staining procedure developed and used by Cora L. Rucker and Arlene F. Koithan of the Brain Research Institute at the University of California at Los Angeles. It is a modification of the Windle, Rhines and Rankin procedure. After the frozen sections are mounted and allowed to dry, they are placed in staining trays and dipped in the various solutions as follows:

Steps	Time (minutes)	Process
1. Ethanol 95%	20 (5)*	Delipidize
2. Chloroform solution	10 (5)	Delipidize
3. Ethanol 95%	2 (2)	Delipidize
4. Ethanol 100%	2	Delipidize
5. Xylene	5	Clear tissue
6. Ethanol 100%	5	Clear tissue
7. Ethanol 95%	2	Clear tissue
8. Ethanol 95%	2	Clear tissue
9. Water (distilled)	2 (2)	Hydrate
10. Staining solution	3–5† (3–5)†	Stain tissue
11. Water (distilled)	2 (4)	Wash off stain
12. Water (distilled)	2	Wash off stain
13. Water (distilled)	2	Wash off stain
14. Formalin acetic solution	4‡ (4)‡	Remove excess stain
15. Water (distilled)	2 (4)	Wash off F.A.S.
16. Water (distilled)	2	Wash off F.A.S.
17. Water (distilled)	2	Wash off F.A.S.
18. Ethanol 95%	2 (4)	Dehydrate
19. Ethanol 95%	2	Dehydrate
20. Ethanol 95%	2	Dehydrate
21. N-Butyl alcohol (100%)	2 (2)	Dehydrate
22. Cedar wood oil	5–10	Clear tissue
23. Xylene	5–10 (5)	Clear tissue
24. Xylene	5–10	Clear tissue
25. Xylene	5–10	Clear tissue
26. Xylene	5–10	Clear tissue
27. Apply cover glass		

*The times given in parentheses indicate the minimum times and reduced number of steps which can be followed for an inferior but suitable rapid staining procedure.

†For fresh solutions, 3 minutes; 5 minutes for old solutions.

‡This time varies, depending upon characteristics of the tissue; close observation will give best results.

TABLE 3-5 PREPARATIONS OF SOLUTIONS FOR THE RUCKER-KOITHAN STAINING PROCEDURE

Buffer solution	
Sodium acetate	7 gm
Glacial acetic acid	2 ml
Water (distilled)	1000 ml
Stock dye solution	
Thionine	1 gm
Water (distilled; boil water before mixing)	100 ml
Staining solution	
Stock dye solution	45 ml
Buffer solution	455 ml
Chloroform solution	
Chloroform	480 ml
Ether	60 ml
Ethanol 95%	60 ml
Formalin acetic solution	
Glacial acetic acid	1 ml
Formaldehyde 39%	1 ml
Water (distilled)	100 ml

TABLE 3-6 FIBER TRACT STAIN

Use of myelin stain for identifying fiber tracts in the brain and spinal cord is the staining procedure developed by Weil for staining the neurilemma of myelinated axons. Alternate sections can be stained for either fiber tracts or cell bodies, resulting in a complete picture of the nuclei and myelinated tracts. The fiber staining technique, however, does not provide much more information, if any, than that obtained by the photographic method.

Steps	Time
1. Potassium dichromate solution	12–24 hours
2. Water (distilled)	2 minutes
3. Water (distilled)	2
4. Water (distilled)	2
5. Ethanol 95%	20
6. Ethanol 100%	10
7. Xylene	5
8. Ethanol 100%	5
9. Ethanol 95%	2
10. Ethanol 95%	2
11. Water (distilled)	2
12. Staining solution	20–30
13. Water (tap, basic), in running water stream	Wash until no precipitate remains in running water
14. Alum solution 4%	Differentiate until black color comes off
15. Alum solution 4%	Differentiate until fibers are barely distinguishable; when glass loses its color, time is about up
16. Water (tap, basic)	2
17. Water (tap, basic)	2
18. Water (tap, basic)	2
19. Potassium ferricyanide solution	Differentiate until fibers are distinguishable
20. Potassium ferricyanide solution	Differentiate completely until they look right under a microscope
21. Water (tap, basic)	2
22. Water (tap, basic)	2
23. Water (tap, basic)	2
24., 25., 26. Ethanol 95%	2, each
27. N-Butyl alcohol	2
28. Cedar wood oil	5–10
29., 30., 31. Xylene	5–10, each
32. Apply cover glass	

TABLE 3–7   PREPARATION OF SOLUTIONS FOR THE WEIL STAIN

Stock staining solution

Hematoxylin	10 gm
Ethanol 95%	100 ml

Note: Be sure to "ripen" this solution for a few hours by leaving it in a 60° C oven with
the lid ajar. After ripening, fill the jar to its original level with 95% ethanol.
Do not use solution until 12 to 72 hours after ripening.

Iron alum solution 4%

Ferric ammonium sulfate, purple	80 gm
Water (distilled)	1920 ml

Potassium ferricyanide solution

Potassium ferricyanide	12.5 gm
Sodium borate	10 gm
Water (distilled)	1000 ml

Staining solution

Stock staining solution	25 ml
Water (distilled)	225 ml
then add	
Iron alum solution 4%	250 ml

Note: Mix only when ready to be used at once. Throw solution away after one use.

Potassium dichromate 4%

Potassium dichromate	4 gm
Water (distilled)	96 ml

TABLE 3–8   BODY TISSUE STAIN

Use of hematoxylin and eosin stain for body tissue is a standard staining procedure for staining non-neural tissues of the body such as the heart, kidney and lungs.

1.	Hematoxylin stain	5 minutes
2.	Water (tap, pH 8–9), running	5
3.	Ethanol 70%	4
4.	Eosin stain 0.5%*	10 seconds*
5.	Ethanol 70%	4 minutes
6.	Ethanol 95%	2
7.	Ethanol 95%	2
8.	Ethanol 95%	2
9.	Ethanol 100%	2
10.	Ethanol 100%	2
11.	Xylene	5
12.	Xylene	5
13.	Xylene	5
14.	Apply cover glass	

Preparation of solutions for the hematoxylin and eosin stain
Hematoxylin stain:

Hematoxylin	1 gm	
Aluminum potassium sulfate	50 gm	Note: Let this solution stand 24
Potassium iodate	0.2 gm	hours before use. This solution
Hydrochloric acid 25%	5 ml	will last only about one week.
Water (distilled)	1000 ml	

Eosin stock stain:

Eosin	2.2 gm
Ethanol 95%	100 ml

Eosin staining solution:

Desired Color of Stain	Staining Solution Concentration	Stock Eosin in cc	95% Ethanol in cc
light	0.1%	23	477
·	0.2%	46	454
·	0.3%	69	431
·	0.4%	92	408
dark	0.5%	115	385

*This staining time depends upon the concentration of the staining solution used and the desired intensity of staining.

Figure 3–48. APPLYING THE COVER GLASS, 1

After the sections are photographed and stained, they must be properly sealed with a cover glass for permanent storage. The last step in the staining procedure is clearing the tissue with xylene. The tissue can be left in this solvent for up to 24 hours before applying the cover glass. Take the slide to be covered out of the xylene and wipe off the back side. Wipe off the xylene along the top edges of the front surface, but do not touch the tissue. Place a small amount of cover glass resin along the right edge of the slide and place the right edge of a cover glass on top of the resin, as illustrated. Be careful not to apply too much resin, and use the thinist type of cover glass (No. 1 thickness). A teaser needle is helpful in applying the cover glass.

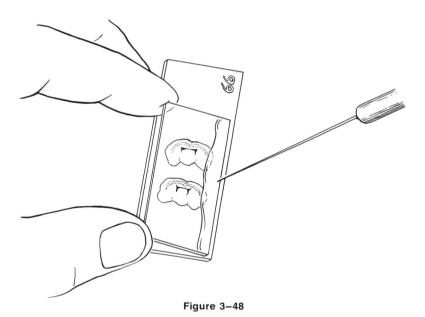

Figure 3–48

Figure 3–49. APPLYING THE COVER GLASS, 2

Next, push the middle of the flexible cover glass down with the teaser needle and run the resin over the tissue, as illustrated. The cover glass should be held between the thumb and index fingers of the left hand (for right-handed persons) so that the left edge of the cover glass can be pushed down close to the slide while it is still being held.

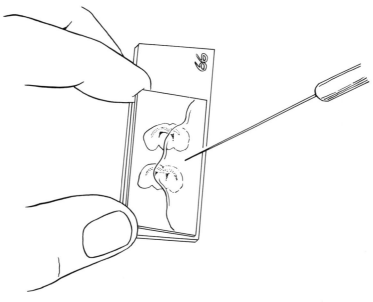

Figure 3—49

Figure 3–50. APPLYING THE COVER GLASS, 3

Next, remove the left thumb from one corner of the glass while the tissue of the index finger still remains between the cover glass and slide. Light pressure from the teaser needle will hold the cover glass while the thumb is withdrawn. Slide the teaser needle toward the left index finger while at the same time rolling the finger over the top of the cover glass to remove it from underneath the cover glass. The resin should spread out evenly over the tissue, forcing all the air out from underneath the cover glass. If air bubbles remain underneath the cover glass, you have not withdrawn your fingers properly. If too much resin has been used, it will ooze out over the edge of the slide. Wipe the excess resin off with a clean, lint-free rag dipped in xylene. Do not touch the clean top surface of the cover glass. If it is desired to repeat the application of the cover glass to the slide because of air bubbles underneath or too much resin running over its top surface, place the slide in xylene and slip the cover glass off laterally.

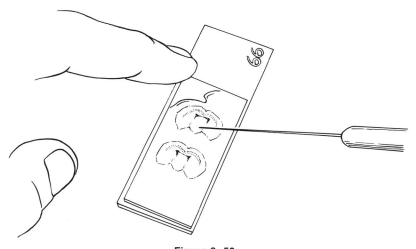

Figure 3–50

REFERENCE

Guzman-Flores, C. M., Alcaraz, M., and Fernandez-Guardiola, A. (1958). Rapid procedure to localize electrodes in experimental neurophysiology. *Bol. Inst. Estud. Med. Biol. (Mexico)* 16:26-31.

CHAPTER 4

BASIC ELECTRONICS

OHM'S LAW

The interrelationships of the variables of a hydraulic mechanism are similar to those of an electrical system but can be described and illustrated more easily. Figure 4-1 shows a water reservoir with four faucets of two different sizes and at two different levels below the water line. For the sake of simplicity in calculation, it is assumed that the water line does not change its height above the faucets no matter how much water flows out of the reservoir; that is, the pressure remains constant while supplying energy to the system. If the height of the water line above the lower faucets is twice that above the upper ones, then the pressure at the lower faucets will be twice that at the upper ones.

The resistance to the flow of water through the faucets is a function of their cross-sectional area and length and the viscosity of the liquid flowing through them. The relationship can be stated $R = \dfrac{V \cdot L}{A}$, where R is the resistance, V is the viscosity of the fluid, L is the length of the faucet, and A is its cross-sectional area.

Current is defined as the quantity of water, Q, that flows through the faucet in a given unit of time ($C = Q/\text{Time}$). When a pressure, P, forces water through a faucet of resistance, R, then a current, C, will flow through the faucet. The equation, $kP = C \cdot R$, shows the relationship between these variables where k is their units constant, or $k = 1$ when the magnitude of each variable is expressed in a standard units terminology.

The four parts of Figure 4-1 show the effects when the value of one variable is changed. If a pressure, P, forces water through a faucet of resistance, R, then a current of water, C, will flow during the time this pressure is applied (Fig. 4-1 A). If the resistance is reduced one half by enlarging the cross-sectional area of the faucet, then twice the current

HYDRAULIC MODEL

Q = Quantity of water

C = Q / time

P = Pressure (i. e., energy)

R = Resistance to water
flow (viscosity, length
and cross-section of faucet)

$$kP = C \cdot R$$

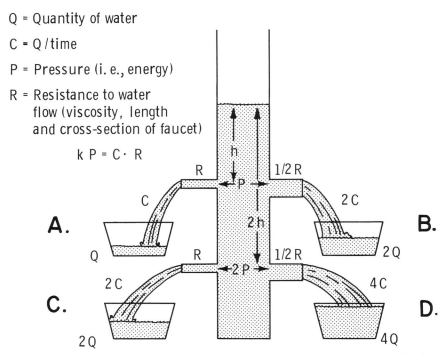

Figure 4–1 Hydraulic model. *A,* A constant pressure, P, is created at the faucet by the water above it at height h. The pressure forces a current of water, C, to flow through the faucet, which has a resistance to the flow, R. The resistance is proportional to the length and the inverse of the diameter of the faucet as well as to the viscosity of the water. The current is defined as the quantity of water, Q, which flows into the bucket in a given time interval. *B,* If the resistance is reduced one half by doubling the diameter of the faucet and if the pressure remains the same, then twice the current will flow. *C,* If the resistance remains at R, but the pressure is doubled by increasing the height of the water above the faucet, then again twice the current will flow. *D,* If P is doubled and R is reduced by one half, then four times the current will flow. The relationship between P, R, and C is expressed by the equation, $kP = C \cdot R$, where *k* is a units constant.

will flow (Fig. 4-1 *B*). If the resistance remains unchanged but the pressure doubles, then again twice the current will flow (Fig. 4-1 *C*). If the pressure is doubled and the resistance is reduced one half, then four times the current will flow (Fig. 4-1 *D*).

An analogous equation, called Ohm's law, expresses a similar relationship between the variables of an electrical system and is written: $kE = IR$, where E is an electromotive force analogous to water pressure, I is a current of unit electrical charges, and R is a resistance to this flow. A unit of electromotive force is defined as the amount of the work performed to bring a unit electrical charge from infinity to the point where it is measured. This procedure is analogous to measuring the work performed when a drop of water is pushed from the outside of the reservoir through a small hole at a place below the water level; it would take more energy to push a drop of water back through a hole at the bottom of the reservoir than one near the water level. For simplicity in computation, it is assumed here also that the source of energy, E, remains constant no matter how many unit charges flow out of its reservoir. A unit charge is equivalent to the charge of an electron if one is discussing electron current, or equivalent to a charge of the same value but opposite polarity if one is discussing conventional current. Electronic diagrams usually refer to the latter, showing current flowing from the plus to the minus pole of a battery (Fig. 4-2).

The resistance to the flow of electrical charges through a resistor is a function of its cross-sectional area, length and resistivity. Resistivity is analogous to viscosity in the hydraulic system. Electrons do not flow through wood as easily as they do through metal because of the higher resistivity of the former. The resistance of a conductor, R, is determined by the equation $R = \dfrac{P \cdot L}{A}$, where P is its resistivity, L is its length, and A is its cross-sectional area. In electrical circuit diagrams, the resistance of a wire is considered to be zero because its resistivity is considered to be zero; therefore, it does not matter how long or thick the wires are.

Current is defined as the quantity of unit charges which flows through the circuit resistance in a given unit of time ($I = Q/\text{Time}$). An electromotive force, E, will cause a current of electrons, I, to flow through a resistance, R. The unit constant k in Ohm's law equation is equal to 1 when E is measured in volts, I in amperes, and R in ohms (Fig. 4-2).

Figure 4-3 shows the effects of changing a variable in an electrical circuit, and is analogous to Figure 4-1, which shows the effects for the hydraulic system. If the resistance in Fig. 4-3 A is reduced one half, then the same electromotive force will move twice as much current through it (Fig. 4-3 B). If the electromotive force is doubled and the resistance remains at the same value, then twice the current will flow (Fig. 4-3 C). If the electromotive force is doubled and the resistance is reduced one half, then the current is increased four times (Fig. 4-3 D).

The concept of a "voltage drop" is very useful in analyzing electronic circuits. When an electromotive force of unknown value causes currents to flow through the branches of a complex electrical circuit, potential differences or voltage drops will appear between the terminals

Figure 4-2 Ohm's law. The battery has a constant electromotive force, E, which produces a current flow of unit electrical charges, I, through a resistance, R. The relationship of these variables is kE = I · R, where k = 1 if E is expressed in units called volts, I in amperes, and R in ohms.

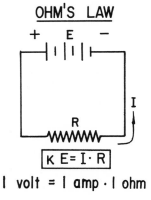

ELECTRICAL MODEL

Q = Quantity of electrons

I = Q / time

$E = \int_p^\infty e^- \, ds$ (i. e., energy)

R = Resistance to electron
flow (conductivity, length
and cross-section of resistor)

$$k E = I \cdot R$$

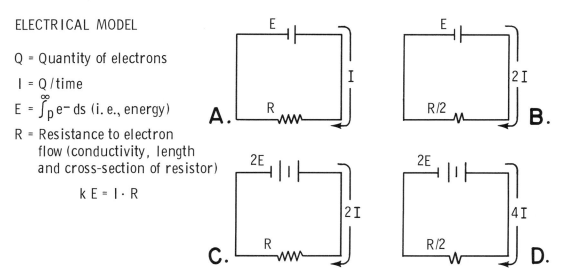

Figure 4–3 Electrical model. *A*, A constant voltage, E, causes a current, I, to flow through a resistance, R. *B*, Reducing the resistance by one half doubles the current flow. *C*, If the resistance remains at R, but the voltage is doubled by adding another battery, then again twice the current will flow. *D*, If E is doubled and R is reduced by one half, then four times the current will flow. The relationship between these variables is expressed by the equation, kE = I · R, which is analogous to the equation, kP = C · R, relating the variables of the hydraulic system.

of each of the resistors in the circuit (Fig. 4-4 *A*). The magnitude of each voltage drop is equal to the product of the current flowing through that particular resistor and its resistance ($e = i_r \cdot r$). One can measure the voltage drop across a resistor and divide that by the value of its resistance to determine the magnitude of the current flowing through it. The value of a resistor is usually written in a color code on its outside, so by measuring the various voltage drops across the resistors and then computing the currents, one can obtain the values of all the Ohm's law variables in the circuit.

An electronic circuit is a circular pathway that unit charges take from a higher energy level of electromotive force to a lower energy level; for example, from one side of a battery to the other. The total of the voltage drops across each resistor connected in series in the circular pathway is equal to the value of the electromotive force of the energy source. Figures 4-4 *B* and *C* illustrate the derivation of this conclusion from Ohm's law. In a *series* circuit, the value of the resistors in the circuit can be added and treated as one resistance, $R = r_1 + r_2$. The same current, *I*, flows through both resistors, so that $e_1 = I \cdot r_1$ and $e_2 = I \cdot r_2$. With these points in mind, it is easy to derive that $E = e_1 + e_2$.

The total current in a *parallel* circuit divides itself between the two parallel branches, as shown in Figure 4-5 *A*: $I_p = i_1 + i_2$. The same electromotive force is exerted on either side of the two resistors: $e_1 = e_2 = E$. Therefore, one can compute, as illustrated in Figure 4-5 *B*, a general formula which states that the total resistance of two parallel branches of the circuit is equal to the product of the resistances divided by their sum.

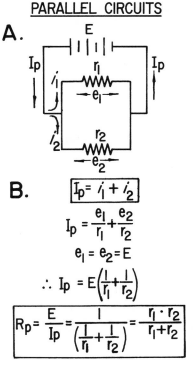

VOLTAGE DROP

A.

B.

C. $E = I \cdot R$
$E = I \cdot (r_1 + r_2)$
$E = I \cdot r_1 + I \cdot r_2$
$$\boxed{E = e_1 + e_2}$$

Figure 4–4 Voltage drop. *A*, A current, I, flowing through a resistance, r, produces a voltage drop between the terminals of the resistor, e_r. *B*, The sum of the voltage drops in the series circuit, e_1 and e_2, is equal to the applied electromotive force, E. *C*, Derivation of proof from Ohm's law. Remember that all resistances in a series circuit are additive.

PARALLEL CIRCUITS

A.

B. $$\boxed{I_p = i_1 + i_2}$$

$$I_p = \frac{e_1}{r_1} + \frac{e_2}{r_2}$$

$$e_1 = e_2 = E$$

$$\therefore \; I_p = E \left(\frac{1}{r_1} + \frac{1}{r_2} \right)$$

$$\boxed{R_p = \frac{E}{I_p} = \frac{1}{\left(\frac{1}{r_1} + \frac{1}{r_2} \right)} = \frac{r_1 \cdot r_2}{r_1 + r_2}}$$

Figure 4–5 Parallel circuits. *A*, Schematic diagram showing total current of parallel circuit, I_p, branching into two smaller currents, i_1 and i_2. *B*, Proof from Ohm's law that the total parallel resistance, R_p, is equal to the product divided by the sum of the parallel resistances, r_1 and r_2. Remember that all currents in a parallel circuit are additive.

COMPLEX CIRCUITS

A.

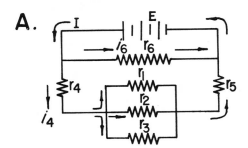

Figure 4-6 Complex circuits. *A,* Schematic diagram showing complex current pathways. *B,* Solution for the total resistance of the circuit, R; see text.

B.

$$R = \frac{r_6 \cdot \left[r_4 + r_5 + \left(\dfrac{r_1 \cdot r_2 \cdot r_3}{r_1 \cdot r_2 + r_1 \cdot r_3 + r_2 \cdot r_3}\right)\right]}{r_6 + \left[r_4 + r_5 + \left(\dfrac{r_1 \cdot r_2 \cdot r_3}{r_1 \cdot r_2 + r_1 \cdot r_3 + r_2 \cdot r_3}\right)\right]}$$

Figure 4-6 illustrates the computations for the total resistance of a complex circuit where both series and parallel current pathways are formed. When three resistances are connected in parallel (Fig. 4-6, r_1, r_2, r_3), first solve for the parallel resistance of two of them $\left(R_{p12} = \dfrac{r_1 \cdot r_2}{r_1 + r_2}\right)$. Then solve for the parallel resistance of the first two with the third $\left(R_{p123} = \dfrac{R_{p12} \cdot r_3}{R_{p12} + r_3} = \dfrac{r_1 \cdot r_2 \cdot r_3}{r_1 r_2 + r_1 r_3 + r_2 r_3}\right)$. The total resistance of a parallel circuit of any number of branches can be seen to be the product of all the resistances divided by the sum of all their cross-products taken two at a time. One of the features of this formula is that the total parallel resistance will always be less than the value of the smallest resistance.

CAPACITANCE

A capacitor is an electronic device that is able to store electrons on one side of its two storage plates and positive unit charges on the other. The degree to which this device is able to store a charge is called capacitance; this is defined as the quantity of charges that can be stored on the plates in order to produce a potential difference of one volt between them, $C = k \cdot \dfrac{Q}{E}$. The unit of capacitance is the farad, and $k = 1$ when Q is in coulombs and E is in volts. The two curves labeled by the number 1 in Figure 4-7 show the voltage drop as a function of time across both a resistor and a capacitor connected in series to a source of electromotive force. The current flowing through the resistor is the same as that flowing through the capacitor, so the voltage drop across the resistor divided by its known fixed resistance is a measure of the current flowing in the circuit. Note that after the switch is closed at

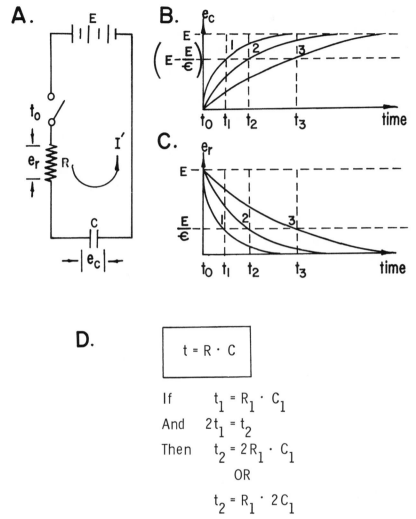

Figure 4–7 Capacitance. *A,* Schematic diagram showing voltage drops across a resistor, e_r, and capacitor, e_c, after a switch in the circuit is closed at time, t_0. *B,* Plot of voltage drop across capacitor, e_c, vs time. The maximum applied electromotive force is E, and the amount of time it takes for e_c to reach $E\,(1 - \frac{1}{\epsilon})$ or $E - \frac{E}{\epsilon}$ is called the time constant, t. For each of the three different curves, 1, 2, 3, the time constants are shown to be different, t_1, t_2, t_3. *C,* Plot of voltage drop across resistor, e_r, vs time. The current going through R is the same as that through C, so the voltage drop, e_c, divided by its fixed resistance, r, is a direct measure of the current through C. Note that the voltage drop across *C* is out of phase with the current going through it. *D,* if the time constant, t_1, is equal to the product of the resistance, R, and capacitance, C, and if the time constant for curve 2 is twice as long as that for curve 1, then either R or C is twice as large in the second circuit as in the first one. The time constant, t, of a circuit is equal to R times C where t is measured in seconds, R in ohms, and C in farads. If R or C is increased, so is the time constant.

time, t_0, the current immediately jumps to a maximum value and then falls off toward zero, while the voltage across the capacitor begins at zero and steadily increases toward the value of the applied electromotive force.

The impedance to the flow of current through a capacitor is different from that for a resistor. A resistor impedes the flow of current by an ohmic process in which the electrons lose energy in the form of heat, but a capacitor impedes the flow by an electrostatic mechanism in which the charges stored on the plates repel the incoming flow of similarly charged particles. The latter does not dissipate any energy. In order to distinguish between the two different types of impedances, the one associated with resistors is called *resistance,* and the one associated with capacitors is called *reactance.* The current through and voltage drop across a resistor are directly proportional to one another, but those for a capacitor are not; the voltage lags behind the current. Because of this lag, resistances and reactances in the same circuit are not algebraically additive. Their combined value, called *impedance,* is equal to the square root of the sum of their squares, $Z = \sqrt{R^2 + X_c^2}$, where each term is expressed in ohms.

At the instant in time when current first begins to flow into a capacitor, it has no reactance, but as time passes and the current flows into the storage plates, a voltage charge begins to build up, causing the reactance to increase until the current eventually stops. When the voltage drop across the capacitor is equal in value to the applied electromotive force, the capacitor is fully charged and has an infinite impedance to further current flow. If the applied voltage is reversed in polarity, the current will flow in the opposite direction, first discharging the capacitor and then recharging it to the opposite polarity. If the polarity is reversed back and forth rapidly so that the capacitor cannot fully charge in either direction, the current will never stop flowing and the reactance will never reach infinity. Similarly, the reactance will never reach the higher values if the capacitance is large enough to enable the alternating currents to flow without fully charging the capacitor before the current is reversed. Thus, the reactance to the flow of alternating currents through a capacitor is inversely proportional to the frequency of current reversal and the capacitance, $X_c = k \cdot \dfrac{1}{f \cdot C}$, where $k = \dfrac{1}{2\pi}$ if f is in cycles per second, C in farads, and the alternating current is a sine wave.

The time that it takes for the voltage drop across the capacitor to reach $(1 - 1/\epsilon)$ times its maximum value, E, is called the *time constant* of the circuit and is directly proportional to the product of the resistance and capacitance, $t = R \cdot C$, where t is measured in seconds, R in ohms and C in farads. Changing the value of either R or C, or both, will change the time constant, as illustrated by the curves labeled 2 and 3 in Figure 4-7. For example, if $t_1 = R \cdot C$ and $t_1 = 2t_2$, then $t_2 = 2 R \cdot C$ where either R or C is doubled in value.

A knowledge of capacitance is important when one is designing and using electronic devices that either stimulate or record voltages from living tissue. For example, electrode tissue interfaces have a fairly large capacitance which affects the wave form of the electrical current through it (Fig. 4-8). Recording amplifiers are often coupled to the tissue through a capacitor in order to prevent the recording of certain low frequency voltages (Fig. 4-10 B). The input capacity of an amplifier determines its limitations for recording the higher frequencies (illustrated in Fig. 4-10 C). An understanding of capacitance is essential for recording and interpreting biologic voltages.

FOURIER ANALYSIS OF WAVE FORMS

Fourier's theorem states that any complex wave form recorded in a finite period of time, for example, an ink-writer record of brain waves, can be reconstructed from a series of sine waves that have integral-multiple frequencies of that period and have varying amplitudes and phase angles. In other words, any biologically generated wave form, no matter how complex, can be broken down into a long series of sine wave components ranging from the low to high frequencies. Figure 4-9 shows the Fourier reconstruction of a rectangular wave form that contains both high and low frequency components. The Fourier series for this particular complex wave is $\frac{A}{1}\sin(f)+\frac{A}{3}\sin(3f)+\frac{A}{5}\sin(5f)\ldots$, where A is the amplitude of the fundamental frequency, and each term represents a sine wave component of decreasing amplitude whose frequency is an odd-numbered integral-multiple of the fundamental frequency, f. In this particular case the phase angles of the sine waves are all the same, which makes the analysis much less complex. Fourier's theorem is important to remember when considering the requirements of a biological amplifier, for it must be able to amplify the high frequency as well as the low frequency components of the biological wave forms.

Amplifiers that are used to record biological potentials must have a broad band frequency response, high input resistance, high common-mode rejection ratio, and high gain with low noise. Each of these requirements will be considered separately.

FREQUENCY RESPONSE

A biological amplifier must be capable of amplifying all the Fourier sine wave components that constitute the biological potential being recorded; that is, it must have a broad band frequency response. Most biological potentials, however, have sine wave components only from zero to approximately 10,000 c/sec. The higher frequency components have such small amplitudes that they become insignificant, in contrast to the lower frequency components that contribute the most to forming the complex wave form. The EEG potentials recorded from large elec-

CAPACITANCE AND RESISTANCE
OF ELECTRODE INTERFACE

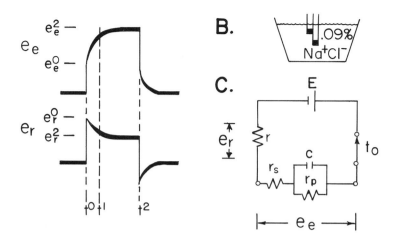

D.

Measured or Known values	r_s	r_p	c
$E = 1.5$ volts	$r_s = \dfrac{e_{r_s}^0}{I^0}$	$r_p = r_{(s+p)} - r_s$	$c = \dfrac{t^1}{r + r_s}$
$r = 10^4$ ohms			
	$I^0 = \dfrac{E}{r + r_s}$	$r_{(s+p)} = \dfrac{e_e^2}{I^2}$	
$t^1 = 3.2 \times 10^{-4}$ sec			
$e_e^2 = 1.15$ volts	$r_s = \dfrac{e_{r_s}^0 \cdot r}{E - e_{r_s}^0}$	$I^2 = \dfrac{e_r^2}{r}$	
$e_e^0 = 0.25$ volts			
$e_r^0 = 1.25$ volts	$e_{r_s}^0 = e_e^0 - 0$	$r_p = \dfrac{e_e^2 \cdot r}{e_r^2} - r_s$	
$e_r^2 = 0.35$ volts			
	$r_s = \dfrac{e_e^0 \cdot r}{E - e_e^0}$	$r_p = 31,000\ \Omega$	$c = 0.029\ \mu F$
	$r_s = 2000\ \Omega$		

Figure 4–8 Capacitance and resistance of electrode interface. *A,* Plot of a voltage drop across an electrode, e_e, and across a known series resistor, e_r, when a known constant voltage, E, is applied from time t_0 to t_2. The instantaneous values of the voltage drops, e_e^0, and e_r^0, are measured at time t_0, while e_e^2 and e_r^2 are the steady-state values recorded at t_2. The time constant of e_e from e_e^0 to e_e^2 is indicated by t_1. *B,* The electrodes for the experiment were insulated stainless steel wire 0.0092 inches in diameter with 0.5 mm tip exposure. They were placed in a solution of physiological saline during the recording of the voltages shown in *A*. The voltage drops across the electrodes would look almost the same if they were located in the brain, an observation which indicates that most of the measured capacitance and resistance of implanted electrodes is that at the electrode-electrolyte interface and not that in the brain. *C,* Schematic diagram of circuit showing a series resistance, r_s, capacitance, C, and parallel resistance, r_p, which are inferred to be located at the interface between the electrodes. When the electrodes are moved from the saline solution into the brain, the values for the resistance and capacitance of the brain tissue are contained in r_s, r_p, and C, which also include the values for the interface elements. *D,* Table

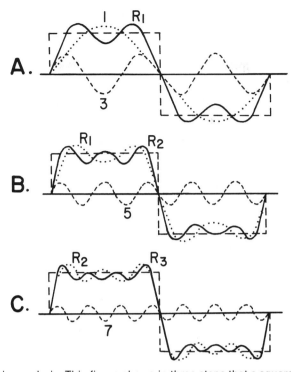

Figure 4–9 Fourier analysis. This figure shows in three steps that a square wave (dashed line) is composed of a series of sine waves. The sine waves in the Fourier series for this particular wave form have decreasing amplitudes and increasing frequencies which are odd-numbered integral-multiples of the fundamental frequency (see text). *A,* The fundamental frequency is shown by the number 1 and the third harmonic by 3. Their resultant when added together is shown by R_1. *B,* R_1 added to the fifth harmonic gives a resultant R_2. *C,* R_2 added to the seventh harmonic gives R_3. Note that the resultants are approaching the wave form of the square wave more exactly as each odd-numbered harmonic in the infinite series is added.

Figure 4–8. *(Legend continued.)*
showing computations of r_s, r_p, and C from known values and values measured from the voltage drops plotted in *A*. When computing r_s, remember that at time t_0 the capacitor is uncharged and shorts out r_p, so that the current I_0, flows only through r and r_s. When computing r_p, remember that at time t_2 the capacitor has an infinite reactance, so that the current, I^2, must flow through r, r_s, and r_p. When computing C, observe that the effective resistance through which C is charged is r and r_s.

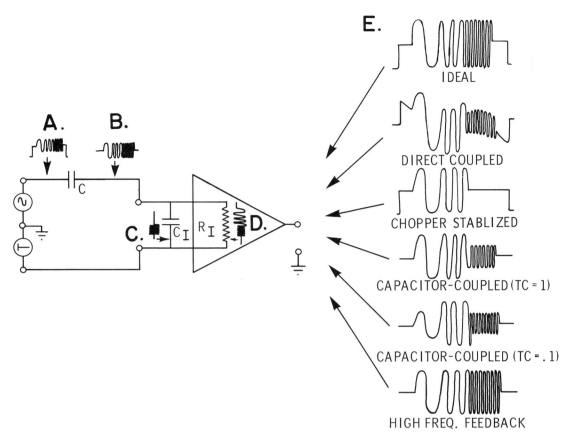

Figure 4–10 Frequency responses of various amplifiers. *A*, The voltage generated by a bio-logic generator consisting of several frequency components: a very low frequency or square wave, low frequency, intermediate frequency and high frequency. *B*, The voltage across a coupling capacitor. Note that the very low frequency component has been eliminated. *C*, The voltage across the input capacitance which is shunting part of the high frequency component. *D*, The voltage across the amplifier input resistance which is missing the capacitor-blocked, very low frequency components and part of the shunted high frequency ones. *E*, Possible wave forms from an amplifier with different input and feedback circuits (see text).

trodes inserted in or on neural tissue have large amplitude components from zero to only a few hundred cycles per second, whereas the extra- and intracellular spike potentials recorded from microelectrodes have large amplitude components ranging to many thousands of cycles per second. Thus, the ideal biological amplifier would be able to amplify all the sine wave components of frequencies from zero to approximately 10,000 c/sec. An even higher frequency response is needed to amplify the spike potentials if it is necessary to know their exact amplitudes.

It is very expensive to construct a high-gain amplifier that has a broad band frequency response to all the components of a biologically generated wave form. Amplifiers that will amplify the zero cycle per second component usually have a high frequency cutoff point at a few hundred cycles per second. Amplifiers designed to amplify the higher frequency sine waves usually will not amplify the very low frequency ones.

The *low frequency* amplifiers are often called direct-coupled amplifiers because each amplification stage is directly coupled to the next without any capacitance inserted between them. One of the problems with such an amplifier is a slow potential drift in the output which is not produced by an input voltage. One way of eliminating this drift problem is to use a "chopper," which produces a higher, chopped frequency from the very low frequency waves, e.g., 400 c/sec. These higher frequency chopped waves are then amplified by a high frequency amplifier which, by an inherent feature, does not drift. The amplified voltage is then reconstituted by the chopper into the original low frequency wave form at the output. The frequencies above the chopper frequency must be cut off with a shunting capacitor to eliminate the chopper noise (Fig. 4-11, HI FREQ. CUT OFF).

The *high frequency* amplifiers are capacitor-coupled between the input and the various amplification stages, thus preventing the zero and very low frequency components of the input voltage from being recorded. The blockade of the low frequency components by a capacitor is the feature mentioned previously that prevents drift in the output voltage. It was also mentioned that a capacitor could not pass a steady current because this current completely charges the plates and causes the capacitor to have an infinite reactance to any further current flow.

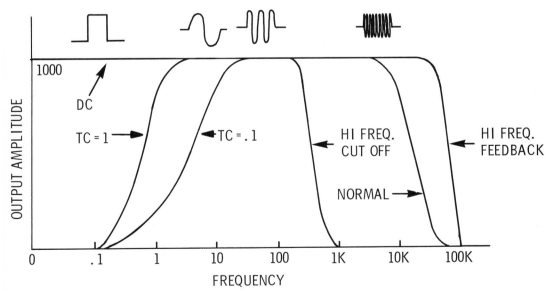

Figure 4–11 Frequency-response curves. These curves show the plot of amplifier output amplitudes vs frequency when a constant-amplitude voltage for each frequency is maintained at the amplifier input. DC: Very low or direct coupled frequencies. TC = 1, TC = .1: Low frequency cutoff for amplifiers with different input time constants, HI FREQ. CUT OFF: High frequency cutoff point for eliminating chopper noise. NORMAL: Normal high frequency amplification range without high frequency cutoff. HI FREQ. FEEDBACK: Extended frequency range with positive feedback and greater amplification for high frequencies.

The capacitors limit the lower frequency response of the amplifier, as well as the drift. It is often desirable to block the lower frequency components in a capacitor-coupled amplifier because large slow potentials caused by a movement of the biological preparation will charge the coupling capacitors to their full capacity and prevent them from passing any current until they have had time to discharge this excess voltage through their draining resistors. The values of the coupling capacitors and their draining resistors determine the time constant of the amplifier. Decreasing the time constant by reducing either the magnitude of the coupling capacitors or their draining resistors will reduce the amplification of these slow blocking potentials, as illustrated in Figures 4-10 and 4-11.

The limitations on the high frequency response of an amplifier arise because small capacitances short out or shunt the signal in the input circuit, preventing all the signal from going through the amplifier (Fig. 4-10 C). These shunting capacitances occur between the recording wires and between the elements in the amplifier. The input capacitance does not shunt the low frequency components because the time constant is so small. Feeding part of the high frequency output voltage of the amplifier back into the input (positive feedback) increases the amplification of the higher frequencies and compensates for their shunted loss at the input.

A recorded biological potential must somehow be displayed visually. An ink-writer works well for writing out low frequency wave forms, but the higher frequency ones require an oscilloscope. The pens of an ink-writer can write out only frequencies lower than 100 to 200 c/sec because their mass and inertia prevent them from keeping up with the higher frequency inputs. An oscilloscope, however, has only to move an electron beam which has a very small mass and inertia (see Fig. 4-12 A), and can respond well to the very high frequency wave forms. The design of a recording system for biological use must take into account both the amplifier and its write-out mechanism, for the weakest link in the recording system determines its overall frequency response.

INPUT RESISTANCE

All amplifiers should have an input resistance that is as high as possible so that most of the generator voltage being recorded is dropped across the amplifier and not other resistances in the input circuit. The voltage that is dropped across the amplifier input resistance is the only part of the generator voltage that is amplified. If the amplifier input resistance is of the same order of magnitude as the other resistances in the input circuit (eg, electrode resistance, the resistance of the generator itself), then a large part of the generator voltage will be dropped across these other resistances and will not be amplified.

Another reason for requiring a high input resistance for an amplifier, especially for a biological amplifier, is to minimize the amount of

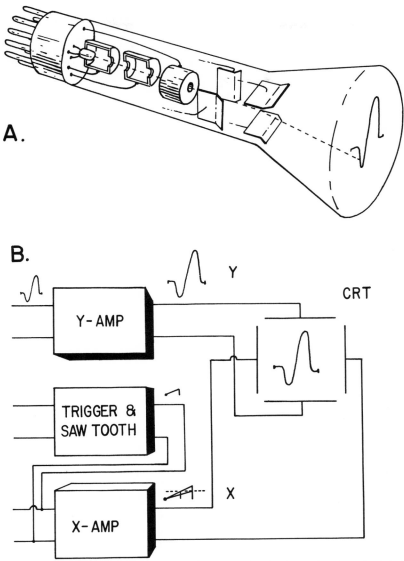

Figure 4–12 Oscilloscope. *A,* Cathode ray tube, CRT, showing horizontal and vertical deflection plates that guide electron beam to phosphorescent tube face. *B,* Block diagram of y-amplifier and linear sweep generator which modulate vertical and horizontal deflection plates of CRT. The linear sweep generator consists of a saw-tooth generator that produces a single saw-tooth wave form when triggered by an external stimulus, and an x-amplifier that determines the horizontal sweep speed by the amplitude of the saw-tooth voltage placed on the deflection plates (larger voltage saw-tooth, faster sweep).

energy or power the generator has to provide to the recording apparatus. An energy drain on a biological system will alter its normal electrical activity, and the recorded voltages will not reflect normal events. The equation $P = I^2 \cdot R$ defines the power, P, that is produced by the generator when a current, I, flows through a total circuit resistance, R. For a constant-voltage generator, when R is increased, I is reduced, making it difficult to relate the power directly to the resistance. By using Ohm's law, the power equation may be rewritten as $P = \dfrac{E^2}{R}$. Since E is a constant, the power drain is decreased by an increase in R. Thus, an infinitely high resistance in the circuit would require an infinitely small amount of power or energy to be expended by the generator while its voltage is being recorded.

Microelectrodes have very large resistances (one million to 100 million ohms) because of the small physical size of their recording tips. It is therefore necessary to have a very large amplifier input resistance when recording from these electrodes. By using a vacuum tube or a transistor in a special input configuration, called a cathode follower or an emitter follower, respectively, a high input resistance can be achieved for the amplifier. However, these devices can provide a resistance of only a few million ohms (megohms), which is not sufficent for many biological recording applications. An electronic amplifying device, called a field-effect transistor (FET), has an extremely high input resistance, ranging from a few megohms to hundreds of thousands of megohms, and is presently the most suitable input device for biological amplifiers.

COMMON-MODE REJECTION

A recording from a biological preparation usually occurs in a laboratory where there are electric motors and electronic devices of many types that must be operated during the experiments. These devices irradiate electric and magnetic fields that induce weak currents in the wires of the input circuit. They are of the same order of magnitude as the weak currents created by the biological generators and are amplified and recorded as unwanted artifacts in the record. A special amplifier, called a *differential amplifier,* eliminates the unwanted signals without affecting the recording of the desired biological signals.

A differential amplifier is essentially two amplifiers whose outputs are connected together so that they subtract from one another. Imagine one amplifier connected to the upward deflection plate of an oscilloscope and the other amplifier connected to the downward deflection plate. If the same signal is applied to both amplifiers, then the oscilloscope beam will not show any signal (Fig. 4-13 A). The differential connection of the amplifiers is analogous to hooking the negative poles of two batteries together and then measuring the potential difference between their positive poles. If the voltage of the batteries is the same value, then the potential difference between them will be zero. If one of the differential amplifier inputs is connected to an electrode recording a biological potential and the other input is connected to an electrode recording no potential (an indifferent electrode), then the biological potential will be amplified and recorded (Fig. 4-13 B). Signals that are applied equally to both inputs or legs of a differential amplifier are rejected from the output (common-mode rejection), but signals that are not the same are amplified and differentially recorded. The signals

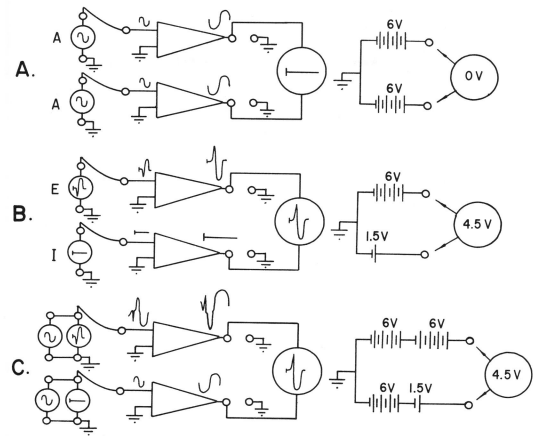

Figure 4–13 Differential amplifier. A differential amplifier is essentially two amplifiers connected together to produce an output which shows the *difference* between the input voltages, such as would occur if one amplifier were connected to the upper vertical deflection plate of a cathode ray tube (CRT) and the other to the lower deflection plate. This is analogous to connecting two batteries back to back and recording their potential difference. *A* shows the results if the same signal (A) is applied to the two amplifiers; *B* shows the results if a different signal is applied to each amplifier, such as that from an active electrode, E, and an indifferent electrode, I, of a biological preparation; *C* shows the effects of *A* and *B* combined, in which the common-mode signal, A (for artifact), is rejected from amplification while the desired biological potential is amplified.

induced in the input circuits by the electric devices in the laboratory are always induced equally in both legs of the amplifier and are therefore rejected, whereas the differences between the biological potentials recorded by each leg are amplified (Fig. 4-13 C).

GAIN AND NOISE

The gain of an amplifier is defined as the magnitude of the output voltage divided by that of the input voltage. It is important to know the approximate magnitude of the biological potential to be recorded in order to select the appropriate gain for the amplifier. It is also important to know how much voltage the amplifier will have to produce in order to operate the write-out instruments. The approximate voltage of the slow-wave electrical activity of the brain, or EEG, is 100 microvolts, and the approximate voltage needed to produce half scale deflection of the beam of an oscilloscope is usually 100 millivolts. Therefore, the gain of the amplifier must be approximately 1000.

There is a lower limit to the magnitude of a generator voltage that can be amplified, no matter what the gain is. With the two inputs shorted together so that zero volts are going into the amplifier, the output will show a slight voltage which is equivalent to a very small input voltage of approximately 5 to 15 microvolts at the input. This output voltage, recorded when the input is shorted, is known as the amplifier noise, and signals smaller than the amplifier noise cannot be distinguished from it. The amplitude of the noise is a function of the frequency response of the amplifier. Narrower band frequency responses result in smaller amounts of amplifier noise being recorded.

CHAPTER 5

CONSTRUCTION OF INTRACRANIAL IMPLANT DEVICES

Most research methods in the neurosciences include either electrical stimulation, recording of biologically generated potentials or functional blockade of the various brain structures. Any type of electrode can be used for both stimulation and recording, and several varieties of subcortical and cortical electrodes are described in the material that follows. There are several methods for performing functional blockade within the brain; three of these methods are presented, one for permanent blockade of cortical structures, a second for permanent blockade of subcortical structures, and a third for performing reversible blockade of both subcortical and cortical tissues. The materials for constructing all the various implant devices can be obtained from supply houses servicing biomedical research institutions; they include insulated stainless steel wire, insulated silver wire, uninsulated tungsten wire, uninsulated platinum wire, uninsulated copper wire, uninsulated constantan wire, Epoxylite wire insulation, potassium nitrite crystals, platinum chloride solution, several sizes of untempered and tempered stainless steel hypodermic-needle tubing, polyethylene tubing, electrical contacts and connector bases, small glass capillary tubing, chemical glassware pipettes and dental cement.

STIMULATION AND RECORDING

SUBCORTICAL ELECTRODES

Various types of electrodes can be constructed, depending upon the requirements for speed and economy in construction and the particular electrophysiological uses intended. The twisted bipolar electrode shown in Figure 5-1 A is small and easily constructed and is generally used for implantation in the brain of the rat. The brains of larger animals require the use of a cannula to guide the small electrode wires into the various structures. This type of electrode, illustrated in Figure 5-1 *B*, has the distinct advantage of permitting implantation without causing trauma or damage to the critical target site. The more rapidly constructed electrode shown in Figure 5-1 *C* is sturdy but requires larger diameter wire which, of course, causes more damage to the recorded or stimulated tissue. Figure 5-1 *D* shows a concentric electrode that will produce a radially symmetrical electrical field when used for stimulation. Figure 5-2 illustrates the construction of the twisted bipolar electrode.

All the electrodes must be made from untempered stainless steel wire; otherwise, the wire cannot be straightened. To straighten the wire, place one end of a 6 inch piece in a vise, grasp the other end with pliers, and then pull on the wire until it stretches about 1 inch. Cut both ends of the wire with a pair of sharp scissors or wire cutters and allow the insulation to shrink back to its natural state and then cut again to the desired length. Diamel insulation is very elastic and does not crack or break when stretched. During construction of the twisted electrode, the weight on the end of the wire will accomplish the stretching and straightening. Before cementing the straightened wires to each other or to the cannulas, scrape the desired amount of insulation from the tips with a sharp scalpel blade. Epoxylite insulation is known to be nontoxic, and in order to protect the brain from toxic metals, this insulation should coat all electrodes, probes and other implant devices except those stimulating and recording surfaces made of stainless steel or platinum.

Figure 5-3 shows a cannula-guided electrode in detail, including an electrode post that is cemented or spot-welded to the electrode in order that it may be held rigidly in the stereotaxic instrument. After implantation this electrode post can be removed with a hot soldering iron if

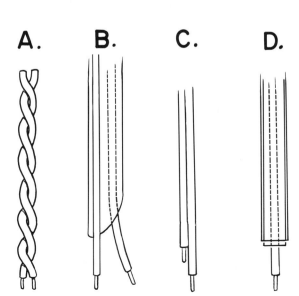

Figure 5–1 Various electrode types. *A*, twisted electrode; *B*, cannula-guided electrode; *C*, parallel electrode; *D*, concentric electrode.

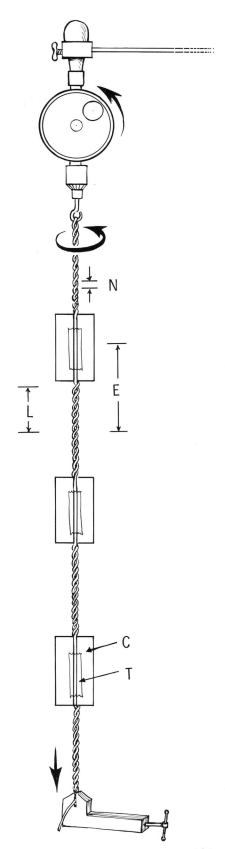

Figure 5–2 Twisted electrode. N, node; L, implant length; E, entire electrode length; C, index card; T, tape.

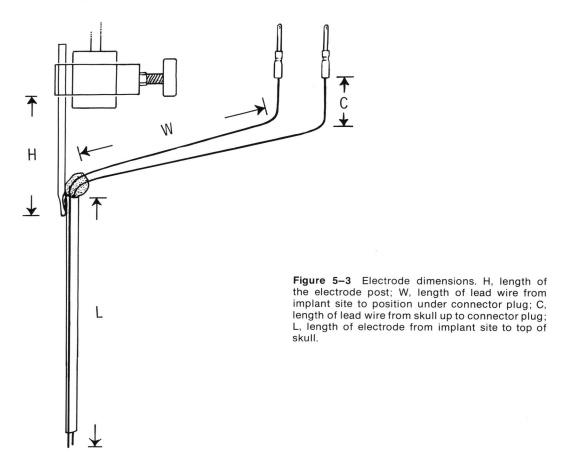

Figure 5–3 Electrode dimensions. H, length of the electrode post; W, length of lead wire from implant site to position under connector plug; C, length of lead wire from skull up to connector plug; L, length of electrode from implant site to top of skull.

cemented with dental cement, or snapped free if spot-welded. The post enables electrodes to be implanted in the brain in close proximity without interfering with one another. Several dimensions of the electrode must be considered before constructing it for a particular use: the implant length of the electrode below the top of the skull, the length of the lead wires running from the electrode to the connector base and the distance the wire must travel from the top of the skull upward to the connector base. Atlases can be consulted to determine the implant length. The distance from the electrode to the connector base must be determined after the site has been chosen for the location of the connector base. The height of the electrode post must be great enough to enable the stereotaxic holder to approach the top of the preparation without touching any previously implanted devices.

CORTICAL ELECTRODES

Figure 5-4 shows a stainless steel skull-screw electrode for monopolar recordings from the surface of the cortex. The screw is placed in an undersize hole drilled in the skull and is driven into the bone while self-tapping its threads. The bottom of the screw should be even with the

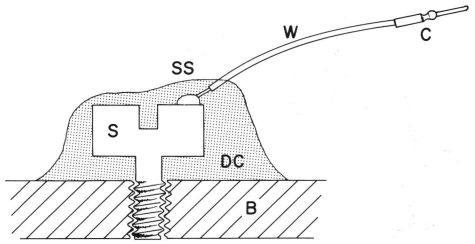

Figure 5–4 Monopolar, stainless steel screw electrode. DC, dental cement; SS, Eutecrod No. 157 silver solder; S, screw; W, lead wire; C, contact; B, skull bone.

underside of the skull so that it will not depress the underlying cortex. The tops of the skull screws should be covered with dental cement to insulate them.

A transcortical recording electrode with platinized platinum recording surfaces is shown in Figure 5-5. This is an ideal cortical electrode because it allows monopolar recording from the surface or the depth of the cortex, or bipolar transcortical recording across the cortex. The platinized platinum recording surface reduces the electrical noise generated by electrolytic processes at the metal-tissue interface and permits the recording of very slow biological potentials that cannot be recorded with stainless steel electrodes. In the latter, the large amplitude, electrolytically generated potentials are very low frequency and are filtered out by the input capacitor of a capacitor-coupled amplifier. Use a No. 10 dental burr to drill the trephine hole in the skull so that underlying dura can be seen clearly. Then insert the electrode in the hole by hand, pushing it through the dura and placing it perpendicular to the surface of the cortex. This electrode is more costly because of the materials and construction time consumed, but it enables great flexibility in recording. It costs about 25 cents for the platinum materials and takes about 30 minutes to construct.

INTRACELLULAR AND EXTRACELLULAR MICROELECTRODES

Figure 5-6 shows an etched metal electrode that is easily constructed for use in recording extracellular spike activity. Tungsten wire can be sharpened electrolytically by repeated dipping in and out of a saturated solution of potassium nitrite while passing current through it. Stainless steel wire can be etched similarly in a solution of hydrochloric acid. Alternating current from a 6.3-volt filament transformer can be used for etching. Tungsten wire is generally preferred because it is very

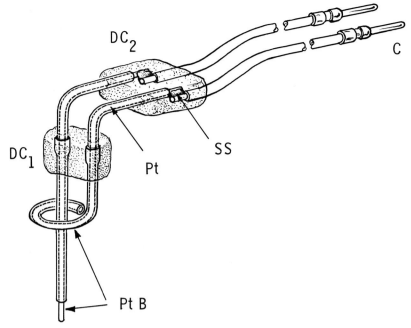

Figure 5–5 Transcortical, platinized platinum electrode. SS, Eutecrod No. 157 silver solder; Pt, platinum wire; PtB, platinum black; DC_1, dental cement holding segments together; DC_2, dental cement encasing solder joints; C, contacts.

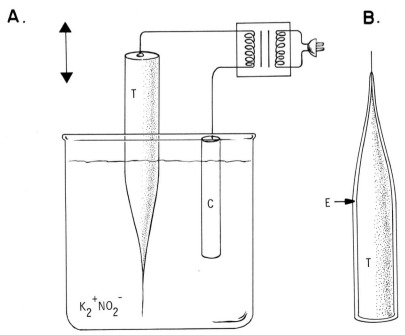

Figure 5–6 Etched metal microelectrode. *A,* Sharpening process: T, tungsten wire, C, carbon rod; $K_2^+NO_2^-$, potassium nitrite. *B,* Completed electrode: T, tungsten wire; E, Epoxylite.

hard. The extreme tip of the wire will always be in the etching solution longer than any other part, as it is the first part to enter and the last to leave. Dipping the wire will thus sharpen it to a needle point, and a taper will develop from the tip to the point of the wire which does not become submerged during dipping. To be suitable for recording extracellular spike potentials, the tip must be less than 1 to 2 microns in diameter, too small to be seen when examined under the highest power of a light microscope. After the metal electrode has been sharpened, it is dipped into an insulating solution, slowly withdrawn, and baked. The cohesive forces of the viscous insulation will cause it to pull back from the sharpened tip of the electrode, thus exposing an uninsulated recording surface.

Figure 5-7 illustrates how a glass micropipette is made. A small diameter Pyrex glass capillary tube is heated to its melting point by an electric coil and at the same time stretched in a straight axial direction. The pull or stretch on the glass tubing must be very straight and uniform or the formed tip of the probe will be curved. Some type of mechanical device must be used to ensure a perfectly straight pull. The magnitude of the heat that melts the glass and the tension that pulls the tubing apart must be continuously variable so that they can be adjusted to the proper values by experimentation. An electric solenoid is generally used to produce the tension of the stretch because the current through it determines the strength of the pull. Examine the pulled electrodes under a microscope. The smooth taper should fade into obscurity because the tip must be smaller than the resolving power of light microscopes. The electrodes are then filled with a conducting ionic solution, usually 0.1 molar potassium chloride, through which the small bioelectric currents can pass from the animal to the recording amplifier. In order to create a noise-free and nonpolarized junction between the ionic solution in the electrode and a silver metal wire connected to the recording apparatus, a silver chloride coat is placed on the outside of the wire before it is inserted into the electrode.

FUNCTIONAL BLOCKADE

Tissue ablations are usually made to create permanent functional blockade in the cortex. Figure 5-8 shows an aspirator and vacuum line trap attached to a vacuum pump that is used to make such ablations. The tip of the aspirator is touched repeatedly to the surface of the cortex through a cut in the pia mater and pressed into the tissue to suck out small amounts. Large trephine holes must be cut in the skull so that the ablation area can be clearly seen. A syringe filled with slightly warm physiological saline must be used to wash away blood that obstructs the view during aspiration. After the ablation is complete, pack the remaining tissue with Gelfoam in order to stop the bleeding.

Several types of lesions can be made in order to produce permanent blockade of subcortical structures: electrolytic lesions, radio-frequency

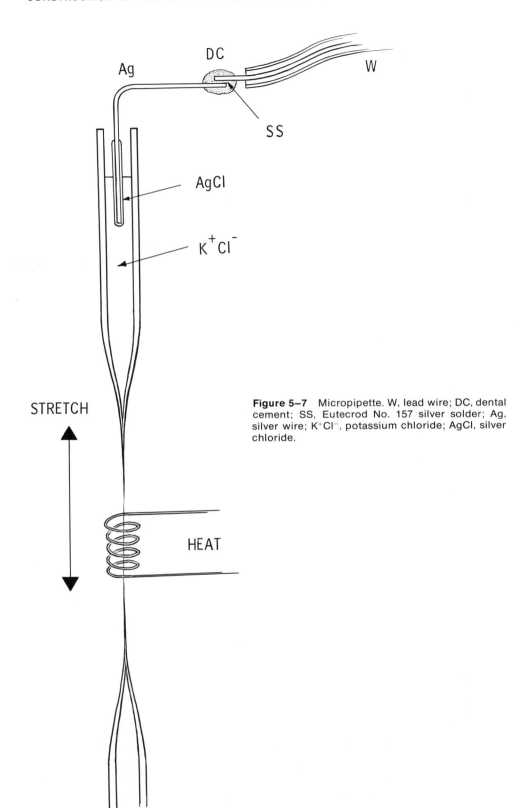

Figure 5-7 Micropipette. W, lead wire; DC, dental cement; SS, Eutecrod No. 157 silver solder; Ag, silver wire; K+Cl−, potassium chloride; AgCl, silver chloride.

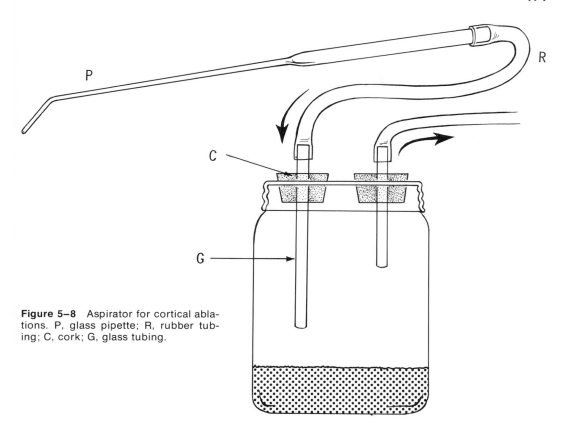

Figure 5–8 Aspirator for cortical ablations. P, glass pipette; R, rubber tubing; C, cork; G, glass tubing.

lesions, heat lesions or freeze lesions. Lesions produced by freezing the tissue are epileptogenic and, furthermore, cause fractures in the frozen areas which sever the blood vessels and can cause subcortical hematomas when the tissue rewarms. The other methods coagulate the tissue as well as the vasculature surrounding the probe tip and thus preclude the possibility of subcortical bleeding. Electrolytic lesions are produced in part by the extreme pH changes at the electrode tip. These lesions are irregular in shape, especially if the current is greater than 3 ma, because of the gas bubble formation at the tip of the electrode. A more uniform lesion is produced by passing anodal current through the lesioning electrode (to an indifferent electrode) than occurs by passing cathodal current. A lesion of 2.0 mm diameter is produced by passing 5.0 ma of current for 30 seconds. Larger or smaller lesions are made by increasing or decreasing the amount of time, leaving the current constant. A 45-volt battery in series with a 5 K ohm potentiometer and 10 ma meter can be used to make the lesions. See Table 5-1 for lesion size vs. current time.

Radio-frequency lesions are produced by heat in the tissue generated from extremely high frequency currents. It is much simpler to control the magnitude of the heat on the surface of a heat probe (Fig. 5-9) than to control the heat from a radio-frequency probe whose current is so easily shunted through non-neural conduction channels. Raising the temperature of the heating element located on the tip of the

TABLE 5-1   LESION SIZE VS. COULOMBS*

Unipolar Electrode
Wire (implanted in brain) connected to positive pole of battery.
Indifferent contact (inserted in anus) connected to negative pole of battery.

Coulombs	Current (ma)	Time (sec)	Diameter of Lesion (mm)
45×10^{-3}	5	15	0.9
150×10^{-3}	5	30	2.0
300×10^{-3}	5	60	2.6

Concentric Electrode
Outer cannula connected to positive pole of battery.

Coulombs	Current (ma)	Time (sec)	Diameter of Lesion (mm)
45×10^{-3}	5	15	1.5
100×10^{-3}	5	20	1.9
300×10^{-3}	5	30	2.3

*Current must be less than 5 ma and time less than 40 to 60 seconds for the lesion size to be predictable from the product of current and time. (From Carpenter and Whittier, 1952.)

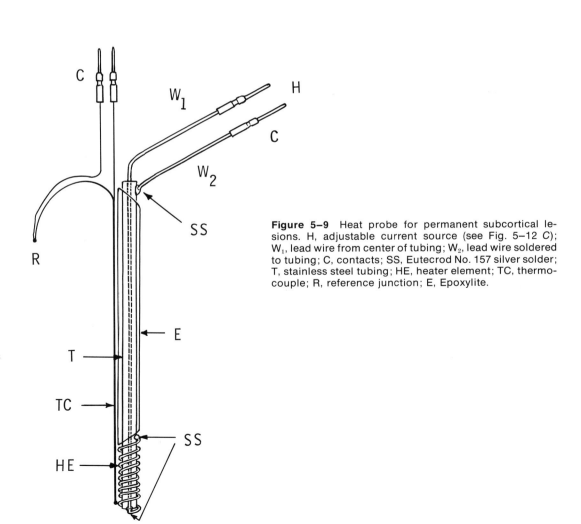

Figure 5-9 Heat probe for permanent subcortical lesions. H, adjustable current source (see Fig. 5-12 C); W_1, lead wire from center of tubing; W_2, lead wire soldered to tubing; C, contacts; SS, Eutecrod No. 157 silver solder; T, stainless steel tubing; HE, heater element; TC, thermocouple; R, reference junction; E, Epoxylite.

heat probe to +10°C above the animal's body temperature for 30 seconds produces little or no significant tissue damage. Raising the temperature to 70°C above the animal's body temperature for 30 seconds creates a lesion about 3 to 4 mm in radius. Other values of lesion size lie within these boundaries and are proportional to the temperature of the surface of the heat probe. The temperature-monitoring and heater-control systems for the heat probe are the same as those used for the cryogenic devices and are illustrated in Figure 5-12.

Reversible functional blockade within the nervous system is generally preferred to the permanent methods because (1) it does not damage the vasculature and cause ischemic-lesion effects to occur downstream from the lesion or ablation site, (2) the size of the functional blockade can be varied to suit the purpose of the particular experiment, and (3) the rapid reversibility of the blocking effects allows the animal to serve as its own experimental control. There are several methods by which neural tissue can be rendered functionally inactive and later returned to its prior condition: injection of various drugs and chemicals into the brain, electrical polarization, local pressure, brief anoxia, heating and cooling. Many of these methods have disadvantages when used to block small, well defined, subcortical brain targets for brief time intervals. The local cooling of subcortical and cortical regions with a cryogenic probe or plate is the most precisely controllable and completely reversible method known at present. If the temperatures on the surfaces of the cooling devices are not reduced below −10°C, the cryogenic blocking process is known to produce no pathologic damage to the tissue and to be completely reversible in its effects.

Figure 5-10 shows the construction of a cryoprobe for use in producing subcortical cryogenic blockade. This probe functions by circulating cold alcohol (methanol or ethanol) through the tip of the probe and by warming the shaft to the animal's body temperature with an electric heater coil. The cooling is thus restricted to the unwrapped portion of the tip. A similar device for producing cortical cryogenic blockade is shown in Figure 5-11. This device, a cryoplate, has a heater to warm the outside surface to the animal's body temperature, thereby enabling its implantation beneath the animal's skin. (The bone must be removed over the cortical region to be cooled.)

The control systems for operating the cryogenic devices are shown in Figure 5-12. To monitor the temperatures of the cryoprobe or cryoplate, thermocouples are attached to their cooled and warmed surfaces. The reference junctions of these thermocouples must be located in a region that remains at a constant temperature during cryogenic blockade. For example, they can be embedded in a mass of dental cement poured into the animal's frontal sinus. When the reference junction and the thermocouple junction attached to the surface of the cryogenic device are at different temperatures, a small voltage will be generated between them which is proportional to the difference in their temperatures ($40\mu/$ °C). This small voltage can be recorded and amplified with a DC amplifier and displayed with a meter calibrated in units of tempera-

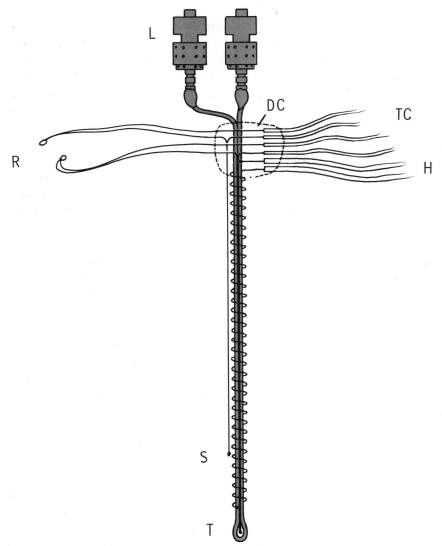

Figure 5–10 Cryoprobe for reversible subcortical cryogenic blockade. L, Luer-Lok connectors; R, reference junctions; TC, thermocouple leads; H, heater leads; DC, dental cement; S, thermocouple attached to shaft of cryoprobe; T, thermocouple soldered to tip of cryoprobe.

ture (Fig. 5-12 *B*). Low resistance heater wires and even lower resistance cable wires must be used so that the heater current can be supplied from a low voltage source. A pure DC voltage source, e.g., an automobile storage battery, must be used for the heater if electrophysiologic recordings are to be made during the cyrogenic blockade. A power transistor is used to control the 3 to 4 amperes of current through the probe, as illustrated in Figure 5-12 *C*. A high resistance meter is used to measure the voltage drop across a very low resistance shunt placed in the emitter circuit of the power transistor. The shunt can be adjusted and the meter calibrated so that the current going through the heater is measured by a meter that is remote to the high current pathway.

The degree of cooling is a function of the flow rate of the cold alcohol through the cryogenic device. The flow rate is determined by the pressure inside the alcohol reservoir tank illustrated in Figure 5-12 *D*. The alcohol, cooled in a dry ice bath, is carried to and away from the implanted cryogenic device by thin polyethylene tubes. These coolant hoses remain flexible during cooling and give complete freedom of movement to the chronic animal preparation. Luer-Lok connectors attach the coolant hoses to the cryogenic devices. Metal male Luer-Lok connectors are soldered to the cryoprobe or cryoplate if repeated attachment and detachment of the coolant hoses are required. However, these metal connectors conduct cold from the alcohol into the dental cement used to secure the cryogenic device to the animal's skull. If plastic female Luer-Lok connectors are used (i.e., a disposable hypodermic needle), then the cold will not be conducted so easily from the alcohol into the dental cement. However, these latter connectors are more easily damaged by the animal and tend to wear with repeated use.

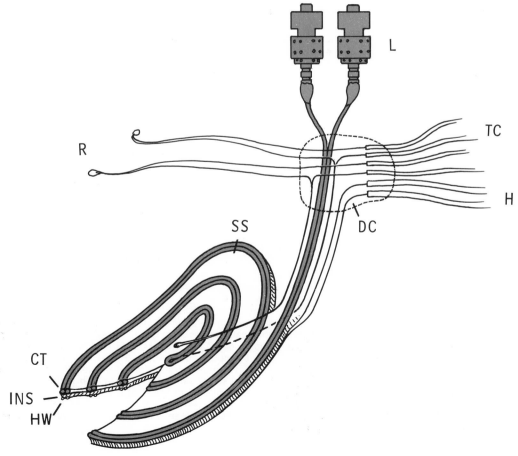

Figure 5–11 Cryoplate for reversible cortical cryogenic blockade. L, Luer-Lok connectors; DC, dental cement; R, reference junctions; TC, thermocouple lead wires; H, heater lead wires; SS, Eutecrod No. 157 silver solder; CT, coolant tube; INS, insulation; HW, heater wire.

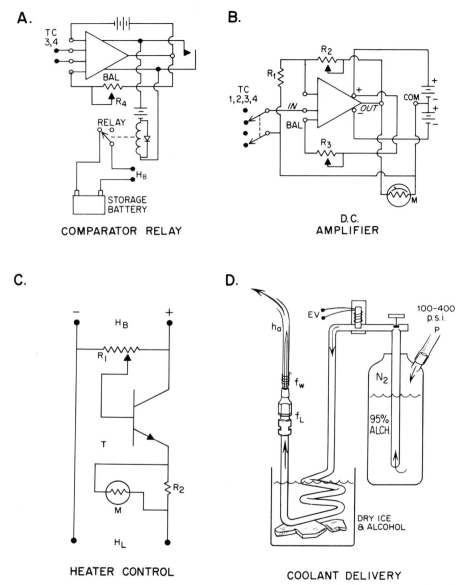

Figure 5–12 Mechanisms for controlling cryogenic system. *A,* Comparator relay: H_B, heater connections; R_4, 25-turn, 50 K ohm Trimpot balance resistor; RELAY, power relay with 4-amp contacts; TC, shaft thermocouple connections. *B,* DC amplifier: M, 2 K ohm, 50 microammeter; R_1, 1 K ohm; R_2, 25-turn, 100 K ohm Trimpot [amplifier gain = $(R_1 + R_2)/R_1$]; R_3, 25-turn, 50 K ohm Trimpot balance resistor. *C,* Heater control: T, power transistor, NPN type, 10-amp current rating; R_1, 1500 ohm potentiometer; R_2, current shunt calibrated for use with meter; M, 2000 ohm, 50 microammeter; H_B, input to heater control system from battery; H_L, output of heater control system to the electrical load of the implant device. *D,* Coolant delivery: EV, electric valve; f_L, Luer-Lok fitting; f_w, wire fitting; h_a, polyethylene coolant hose, PE100; P, rubber pressure hose from a variable pressure regulator.

If the animal bites through the coolant hose, or the alcohol circulation is stopped for some other reason, then the cooling will cease but the heating will continue and cause the temperature of the shaft to become hot enough to create permanent destruction of the brain. A comparator relay system will prevent such a catastrophe by turning the heater current off if the shaft temperature should ever rise above the animal's body temperature. This comparator relay circuit constantly monitors the temperature of the shaft thermocouple and controls the voltage source of the heater circuit, as illustrated in Figure 5-12 A.

Once the cooling and heating adjustments have been made, the cryogenic devices will remain at constant temperatures. When the surface of the cryoprobe tip is above 25°C, the tissue immediately adjacent to the probe is not functionally blocked (Skinner and Lindsley, 1967). Lowering the tip temperature to 0°C produces a functional blockade that extends approximately 3 mm from the surface of the probe (Skinner and Lindsley, 1968). Temperature measurements have been made within the brain which confirm the restricted functional blockade produced by such local cooling (Dondey et al., 1962).

Figure 5-13 shows a simple cryogenic system which does not require complicated control systems and can be used to demonstrate the effects of reversible cryogenic blockade in the cortex of the rat brain. By using standard-size components in the system, a carefully monitored

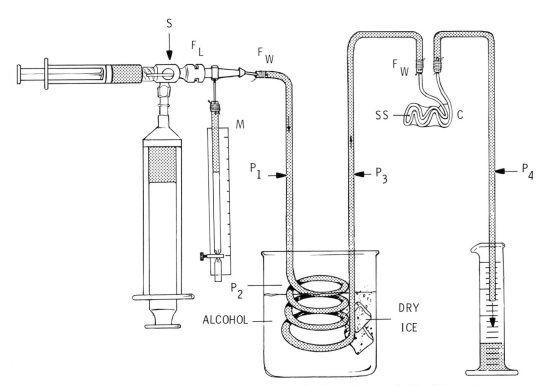

Figure 5–13 A simple cryogenic system. S, stopcock; F_L, fitting, Luer-Lok; F_W, fitting, wire; M, manometer; P_1, P_2, P_3, P_4, polyethylene tubing; SS, silver solder; C, cryoplate.

flow rate or coolant pressure will be an accurate calibrated measure of the temperature of the surface of the cryogenic device, thus eliminating the need for a complicated electronic temperature-monitoring system. If ethyl alcohol is used as the coolant, the greatest force that one can exert on the syringe will not produce a temperature on the surface of the implanted cryogenic device that is below −10°C. If methyl alcohol is used as the coolant, a smaller force on the syringe will be sufficient to produce the same amount of cooling, and the flow rate or coolant pressure must be monitored very accurately to prevent permanent damage by freezing. (See Table 5–2.)

CHEMOSTIMULATION

A device for inserting crystalline chemical compounds into small subcortical regions of the brain is illustrated in Figure 5-14. The inner cannula of this chemoprobe is filled with powdered crystals of the desired chemical-stimulating substance and is then inserted into the

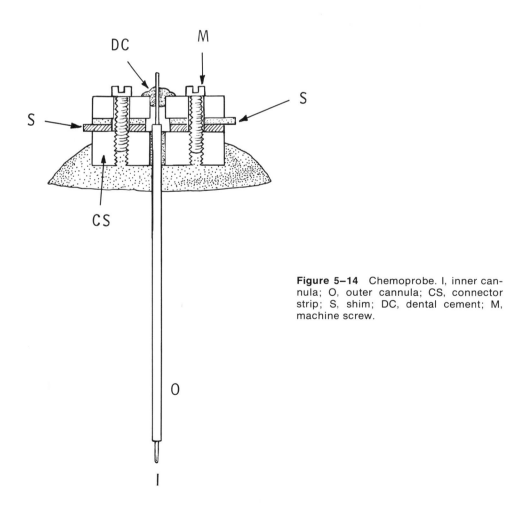

Figure 5–14 Chemoprobe. I, inner cannula; O, outer cannula; CS, connector strip; S, shim; DC, dental cement; M, machine screw.

outer cannula which has been previously implanted stereotaxically. Small shims are placed between the holders of the two cannulas as illustrated. These are removed one at a time in order to permit incremental penetrations of the inner cannula into deeper regions of the brain.

PROCEDURES FOR CONSTRUCTION OF VARIOUS IMPLANT DEVICES

Detailed outlines are given in the material that follows for steps in construction of various implant devices. The letters enclosed in parentheses refer to the letters labeling the figure indicated in the heading for each implant device. All the dimensions and specifications for the materials mentioned are those used by the manufacturers, and at the end of this section the names and addresses of the manufacturers or distributors are given, along with more detailed descriptions for each type of material.

TWISTED ELECTRODES (FIGS. 5–1 A AND 5–2)

1. Stretch a 6 to 8 foot length of stainless steel wire between a 3 to 4 pound vise and a mounted drill, as illustrated in Figure 5-2. Loop the wire over a hook in the drill and clamp both ends in the vise, letting the vise hang on the wires to exert a tension.

2. Tape pieces of index cards (T, C) between the two wires at intervals determined by the implant length of the electrode (L).

3. Move the taped cards up and down the length of the wire until the distance beween them is twice the implant length. One entire electrode length is designated by E.

4. Twist the wire with the drill slowly until the nodes (N) that form are approximately 1 mm apart.

5. Cut the wire with a pair of sharp scissors or wire cutters into individual electrodes and scrape 0.5 mm of insulation off each tip with a sharp scalpel blade.

6. Scrape 3 mm of insulation off the leads of the electrode and crimp or solder contacts onto them.

Materials: stainless steel wire, nontempered, 0.0048 inch diameter, Diamel insulated (use 0.0092 inch diameter wire for brains larger than a rat brain); index cards; tape; contacts.

CANNULA-GUIDED ELECTRODE (FIGS. 5–1 B AND 5–3)

1. Cut a piece of tempered, 26-gauge tubing to a length that is 3.5 mm shorter than the desired implant length. Allow 3 to 4 mm to extend above the skull after implantation. Grind the end of the tubing smooth, using a small cut-off disk. Ream out the inside edges with a hypodermic needle.

2. Stretch and straighten a segment of insulated, 0.0048 inch diameter, stainless steel wire that is long enough to reach from the desired implant site all the way to its position in the connector plug.

3. Scrape 0.5 mm of insulation from one end of the wire. Apply Epoxylite to the wire and to the tubing and allow to dry a few seconds until tacky. Stick the wire to the outside of the cannula, extending the scraped end 3.5 mm beyond the end of the tube. Be careful not to smear insulation onto the scraped end of the wire.

4. Bake in a 150°C oven for a few seconds and then check to see that the wire is in its proper position. If so, bake for an additional 10 minutes. Apply a second coat of Epoxylite and bake at least 30 minutes. Do not allow the Epoxylite to get in the tube openings.

5. Stretch and scrape another segment of stainless steel wire as in steps 2 and 3. Lower the wire through the cannula until its scraped end is even with that of the wire cemented to the cannula. Make a bend in the inner wire at the point where it emerges at the upper end of the tube.

6. Cut the lead wires to the desired length, scrape 3 mm of insulation off the ends and attach the contacts.

7. Spot-weld or cement on a piece of untempered stainless steel tubing to make an electrode post (Fig. 5–3, *H*). Note: The inner wire should be removed from the electrode during implantation and then afterward pushed through the cannula into the tissue and cemented in place.

Materials: stainless steel wire, untempered, 0.0048 inch diameter, Diamel insulated; stainless steel hypodermic-needle tubing, tempered, 26-gauge, 0.018 inch outside diameter, 0.00425 inch wall thickness; Epoxylite; contacts.

PARALLEL ELECTRODE (FIG. 5–1 C)

1. Stretch and cut to the desired length two segments of insulated, 0.0092 inch diameter, stainless steel wire.

2. Scrape 0.5 mm of insulation from the end of the wires.

3. Cement the wires together and adjust the separation between their tips to the desired length.

4. Apply Epoxylite and bake at 150°C for 15 minutes. Be careful not to cover the uninsulated surfaces. Apply a second coat of Epoxylite and bake for 30 minutes.

5. Scrape 3 mm of insulation off the leads and apply the contacts.

Materials: stainless steel wire, untempered, 0.0092 inch diameter, Diamel insulated; Epoxylite; contacts.

CONCENTRIC ELECTRODE (FIG. 5–1 D)

1. Cut a segment of tempered, 26-gauge, stainless steel tubing to the desired length, making it 3.5 mm shorter than the desired implant length and allowing 3 to 4 mm to protrude above the skull after implan-

tation. Grind the ends smooth and ream the inside edges with a hypodermic needle.

2. Insert a segment of insulated, 0.0048 inch diameter, stainless steel wire into the cannula, its length being determined by the implant and lead lengths. Allow exactly 3.5 mm of this inner wire to extend beyond the end of the cannula. Bend the upper part of the inner wire over the upper edge of the cannula so that it will remain in position.

3. Dip the entire tube and wire in Epoxylite, withdrawing slowly, and bake 10 minutes at 150°C. Apply a second coat of insulation and bake for at least 30 minutes.

4. Scrape 0.5 mm of insulation off the tip of the inner wire.

5. Scrape 0.5 mm of insulation off the tip of the cannula (see Fig. 5-1 D).

6. Remove the insulation from the upper edge of the cannula, and solder a lead wire to the cannula using Eutecrod No. 157 solder and flux.

7. Cut the lead wires to the desired length and scrape 3 mm of insulation from their ends and attach contacts.

8. Attach an electrode post as seen in Figure 5-3.

Materials: stainless steel hypodermic-needle tubing, tempered, 26-gauge, 0.018 inch outside diameter × 0.00425 inch wall thickness; stainless steel wire, untempered, 0.0048 inch diameter, Diamel insulated; Epoxylite; Eutecrod No. 157 solder and flux; contacts.

MONOPOLAR, STAINLESS STEEL SCREW ELECTRODE (FIG. 5–4)

1. Using Eutecrod No. 157 solder and flux (SS), attach a lead wire (W) to a stainless steel machine screw (S). The upper edge of the screw should be ground to a rough surface before the flux and solder are applied. The lead wire can be either the Diamel-insulated, stainless steel wire or any polyvinyl-insulated, small-gauge, multistranded copper wire. This wire will be encapsulated inside the dental cement during implantation.

2. Remove all excess flux by soaking the screws in several different alcohol-rinse baths.

3. Cut lead wires to the desired length and apply contacts (C).

Materials: stainless steel machine screws, No. 0-80 × $^3/_{16}$ inch, fillister head (for use in rat brains); stainless steel machine screws, No. 2-56 × $^3/_{16}$ inch, fillister head (for use in larger animals); Eutecrod No. 157 solder and flux; lead wire; contacts.

TRANSCORTICAL, PLATINIZED PLATINUM ELECTRODE (FIG. 5–5)

1. Cut two segments of platinum wire (Pt) into 12.5 mm lengths and scrape off the insulation, if any. In one segment bend a small loop and then bend the loop perpendicular to the stem.

2. Dip in Epoxylite and remove any excess insulation from the loop so that the opening remains clear. Bake at 150°C for 30 minutes. Repeat

dipping and baking for a total of three coats. Bake the last coat for one hour.

3. Scrape 0.5 mm of the Epoxylite from the end of the unbent segment and, on a cut-off disk or fine carborundum paper, grind the Epoxylite off the bottom of the loop.

4. Put the straight segment through the loop, allowing it to protrude 3 mm. Join the two segments with dental cement (DC_1).

5. Bend the sections of wire above the cement about 80 degrees from the original axis.

6. Using Eutecrod No. 157 solder and flux (SS), attach lead wires to the platinum segments. Determine the length of the leads according to the positions of the implant site and the connector plug.

7. Encase the solder joints in dental cement (DC_2).

8. Cut lead wires to desired length and attach contacts (C).

9. Platinize the loop and tip as follows: Connect in series the electrode, a 1.5-volt battery, an ammeter, a 1500-ohm potentiometer and a platinum coil. Attach the electrode to the negative pole of the battery. Dip the electrode and the coil in a platinum-chloride solution and adjust the potentiometer until approximately 1.5 ma of current are flowing. Lift the electrode out of the solution every few seconds to dispel any bubbles that form. Allow 60 to 90 seconds for thorough coating; the electroplated surfaces (PtB) of the tip and loop should turn a dull gray-black color.

Materials: platinum wire, 90 per cent platinum and 10 per cent iridium, 0.007 inch diameter; Epoxylite; lead wire, small-gauge, stranded, polyvinyl-insulated (use two different colors to allow loop and tip to be color-coded); dental cement; platinum-chloride solution, 5 parts 5 per cent platinum chloride to 1 part hydrochloric acid and 4 parts water, mixed with a trace of lead acetate; Eutecrod No. 157 solder and flux; contacts.

ETCHED METAL MICROELECTRODE (FIG. 5–6)

1. Cut and straighten a 3-inch length of tungsten wire.

2. Connect the tungsten (T) and a piece of carbon rod (C; e g, the inside part of a pen-cell battery) to a filament transformer.

3. Place the carbon rod in a saturated solution of potassium nitrite, and dip the tungsten in and out of the solution until a proper taper and sharpness are obtained. The tip should be so small that it cannot be seen under the highest power of a light microscope.

4. Dip the electrode in Epoxylite, and pull it out slowly to make a uniform coat. Bake at 150°C for 30 minutes. Apply a second coat and bake for one hour. The finished electrode (T), seen in Figure 5-6 B, will be fully coated with Epoxylite (E) except for the bare recording tip.

5. Attach a contact directly to the wire.

Materials: tungsten wire, approximately 0.015 inch diameter; carbon rod; potassium-nitrite crystals; Epoxylite; contacts.

MICROPIPETTE (FIG. 5–7)

1. Use Pyrex glass capillary tubing with 1.5 mm outer diameter.

2. Put the tubing through a heater coil and stretching device (i e, a micropipette puller). The stretch must be linear and uniform to create a straight tip. Adjust the heat and stretching tension to produce the desired taper and tip size. The tip must be invisible under the highest power of magnification of a light microscope.

3. Place several pulled micropipettes in a holder and set them in a jar of methyl alcohol.

4. Heat the methyl alcohol to about 40°C for one hour. Caution: Do not use a flame; use an electric oven.

5. Immediately place the warm jar in a vacuum chamber for 8 minutes. Afterward, release the vacuum slowly.

6. Remove the alcohol by setting the micropipettes in distilled water for three minutes.

7. Place the micropipettes in a jar containing the potassium-chloride solution. The level of the ionic solution should cover them. Allow the pipettes to stand in the solution for three days in order to fill the tips with the ionic solution by diffusion. Use the filled micropipettes within five to 10 days.

8. Insert a silver wire (Ag) into the solution (K^+Cl^-) in the pipette for a lead wire; the lead wire should be previously electroplated with silver chloride (AgCl).

9. Using Eutecrod No. 157 flux and solder (SS), attach a hook-up wire (W) to the silver lead wire and cover the solder joint with dental cement (DC).

10. The impedance of the microelectrode must be at least 25 to 50 megohms in order for the cross-sectional area of the tip to be small enough (less than 0.5 microns) to impale a neuron without rupturing it.

Materials: Pyrex glass capillary tubing, 1.5 mm outer diameter; potassium-chloride solution, 0.1 molar, made slightly acidic by adding a few drops of hydrochloric acid to the 1 to 2 liters of solution used to fill the electrodes (it may be saturated with methylene blue dye in order to make it more visible); methyl alcohol; distilled water; silver wire; Eutecrod No. 157 solder and flux; dental cement; stranded hook-up wire.

CORTICAL ABLATION ASPIRATOR (FIG. 5–8)

1. Heat a standard chemical glassware pipette (P) in a flame to cause a bend about 1.5 cm from the end. The tip should be 1.5 to 3.0 mm diameter and as thin-walled as possible.

2. Using a flexible rubber vacuum hose (R), attach the pipette to the glass tube (G) of a vacuum trap.

3. To make the vacuum trap, run long and short glass tubes through corks (C) that are fitted into the lid of a jar.

4. Using another piece of rubber vacuum hose, connect the shorter glass tube in the trap to a vacuum pump.

Materials: standard chemical glassware pipettes; a jar with tight-fitting lid; 2 corks with 0.25 inch holes; one foot of $1/4$ inch glass tubing; 10 to 20 feet of 0.25 inch rubber vacuum hose.

HEAT-LESION PROBE (FIG. 5–9)

1. Cut a piece of 26-gauge, tempered, stainless steel tubing (T), its length determined by the implant depth plus 3 to 4 mm to protrude above the skull after implantation.

2. Using Eutecrod No. 157 silver solder and flux (SS), attach a small diameter, insulated, silver heater wire (H) several millimeters above one end of the tube, a distance which depends upon the desired length of the heating element. (In the remainder of this section, solder always refers to Eutecrod No. 157.) Wrap the wire toward the distal tip of the tube in spirals which are less than 0.5 mm apart.

3. Cement the heater coil in place with a single coat of Epoxylite baked at 150°C for 15 minutes.

4. Insert a piece of insulated, 0.0048 inch diameter, stainless steel wire (W_1) through the tube, scrape off 0.5 mm of insulation, and solder it to the distal end of the heater wire. Push the stainless steel wire out of the end of the cannula so that the soldered junction does not short out to the stainless steel tube. Apply Epoxylite to this region and bake at 150°C for 30 minutes.

5. Solder a piece of any type of lead wire (W_2) to the top of the tube.

6. Cement a thermocouple (TC) to the outside of the cannula. The thermocouple is constructed from a piece of constantan wire approximately 80 mm long and two pieces of copper wire each 40 mm long plus the length of lead wire required. The first step is to stretch and straighten the thin thermocouple wires. Then dip them in Epoxylite and bake at 150°C for 10 minutes. Repeat the dipping and baking for a total of three coats, but bake the last coat for at least 30 minutes. Cut the wires to their proper length and then scrape 1 mm of insulation from the ends of the constantan wire and from one end of the copper wires. Place one copper wire and one constantan wire side by side with their bare, scraped ends adjacent to each other. Solder these wires together, and then rinse off the excess flux with alcohol. Place the soldered junction on the most distal portion of the heating element. Using Epoxylite, cement the thermocouple to the shaft of the probe. Bake at 150°C for at least 15 minutes. Next, solder the other copper wire to the free end of the constantan, thus forming the reference junction (R). Rinse off the excess flux and then insulate the reference junction with Epoxylite or dental cement.

7. Insulate the section of tubing above the heating coil with five baked coats of Epoxylite (E). Do not put any Epoxylite on the heater coil.

8. Secure the heater wire with dental cement as it emerges from the upper part of the cannula, thus protecting the lead from scraping on the upper edge of the cannula and forming an electrical short.

9. Attach the electrode post as illustrated in Figure 5-3.

10. Cut the lead wires to their proper lengths, scrape 3 mm of insulation from their tips, and apply the contacts (C).

Materials: stainless steel hypodermic-needle tubing, tempered, 26-gauge, 0.018 inch outside diameter × 0.00425 inch wall thickness; silver wire, 0.003 inch diameter, insulated; 0.005 inch diameter copper wire, uninsulated; 0.005 inch diameter constantan wire, uninsulated; Eutecrod No. 157 silver solder and flux; Epoxylite; any type of lead wire; contacts.

CRYOPROBE (FIG. 5–10)

1. Cut a piece of untempered, 26-gauge, stainless steel tubing approximately 115 mm in length, depending upon the desired implant depth. Straighten the tubing with pliers and then bend it in the middle in a tight U turn. Both sides of the U tube should be touching except for a slight separation at the bend. Ream out the inside edges of the tube openings and force alcohol through the tube with a syringe equipped with polyethylene tubing and a wire fitting (see f_w in Fig. 5-12) to be certain that the tube is unplugged and has not been pinched off by the bend. Make sure that the U tube is straight and untwisted and then clean its outside with alcohol.

2. Using Eutecrod No. 157 silver solder and flux, solder the two sides of the U tube together, forming a solder joint from the bent portion of the tube up to the desired implant length of the probe. (For the remainder of this and all following sections, solder refers to Eutecrod No. 157 solder and flux.)

3. Remove the excess flux with alcohol swabs and then soak the U tube for a few minutes in xylene followed by alcohol. Dip the soldered length of the probe in Epoxylite and bake at 150°C for 15 minutes. Repeat the Epoxylite dipping and the baking for a total of six coats. Bake the last coat for an hour or until the Epoxylite has turned a dark brown color.

4. Remove insulation from the tip to form the desired length of cooling surface.

5. Solder a small diameter (0.003 inch), insulated, silver wire approximately 10 inches long to the part of the cooling surface that is nearest the beginning of the insulation on the shaft.

6. Construct a thermocouple from a piece of constantan wire approximately 80 mm long and two pieces of copper wire each 40 mm long plus the amount of lead wire needed. Stretch and straighten the wires, and then dip them in Epoxylite, drawing them out very slowly in order to make a smooth and uniform coat of insulation. Bake at 150°C for 10 minutes. Repeat the dipping and baking for a total of three coats, but bake the last coat for at least 30 minutes. Cut the wires to their proper lengths and then scrape 1 mm of insulation from both ends of the constantan wire and from one end of the copper wires. Place one copper wire adjacent to the constantan wire with their bare tips next to each

other. Solder these tips together using plenty of flux and without burning the insulation off the unscraped parts of the wires.

7. Lay the thermocouple wires in the groove on one of the soldered sides of the U tube. Using plenty of flux, solder the thermocouple junction to the most distal tip of the cooling surface (T). Again use care not to burn the insulation off the wires. Wash off the excess flux with alcohol squirted from a syringe with a 25-gauge needle. Using Epoxylite, cement the thermocouple wires to the shaft of the U tube and bake at 150°C for 15 minutes.

8. Wrap the previously attached heater wire over the tip thermocouple and up the shaft, making each loop less than 0.5 mm apart. Continue the wrapping for 5 mm above the point on the probe that will emerge above the skull after implantation. Attach a small clip to the remaining length of heater wire in order to maintain a tension on the wrapped coil while it is being cemented in place. Apply a coat of Epoxylite to the coil with a wooden stick and bake at 150°C for 15 minutes. Do not apply Epoxylite to the cooling surface.

9. Cement a copper and constantan thermocouple to the shaft of the probe with the junction about 5 mm above the cooling surface (S). Construct the thermocouple junction as described in step 6.

10. Solder the other prepared copper wires to the unattached, scraped ends of the constantan wires to form the reference junctions (R). Rinse off the excess flux with alcohol and encapsulate the reference junctions in dental cement.

11. The lead wires must be soldered within 2 mm of the point where the heater wire leaves the coolant tubes or overheating will occur during operation and melt the small diameter wire. Use plenty of flux to solder the lead wire in place, and then wash off the excess flux with alcohol squirted from a syringe with a 25-gauge needle. Absolutely all the flux must be removed because it is electrically conductive and will short out the heater circuit.

12. Solder a second lead wire for the heater to the upper part of the U tube to carry the return current from the coil. Use a heat sink to protect the other solder joints.

13. Encapsulate the bare solder joints of the lead wires in dental cement (DC).

14. Cut the lead wires to their proper lengths (TC, H) and apply contacts.

15. Using an ohmmeter, check for shorts between all the thermocouple and heater wires. (Use R × 100 K scale.) On the R × 1 scale of the ohmmeter, the heater wires should measure approximately 0.5 ohms, and the thermocouples should measure approximately 3.5 ohms.

16. If the male metal Luer-Loks (L) are used to connect the cryoprobe to the coolant hoses, drill small holes in them so that the dental cement will adhere to them. Then solder these connectors to both ends of the metal coolant tube. Use a heat sink to avoid melting the dental cement encapsulating the large diameter, heater lead wires.

Materials: stainless steel hypodermic-needle tubing, untempered, 26-gauge, 0.018 inch outside diameter × 0.00425 inch wall thickness;

Eutecrod No. 157 silver solder and flux; Epoxylite; silver wire, 0.003 inch diameter, insulated with a 0.001 inch coat; copper wire, 0.005 inch diameter, uninsulated; constantan wire, 0.005 inch diameter, uninsulated; stranded hook-up wire, 22-gauge diameter, insulated; Luer-Lok connectors; contacts.

CRYOPLATE (FIG. 5–11)

1. Cut and straighten a long segment of untempered, 26-gauge, stainless steel tubing to the length required to make the size cryoplate desired. Bend the tubing in the middle into a long U tube and then bend these parallel tubes into a spiral (CT) the size and shape of the cortical structure to be cryogenically blocked.

2. Apply Eutecrod No. 157 solder to form a sheet of metal between the spiral loops (SS).

3. Coat the outer surface of the cryoplate (the side which will not be in contact with the brain) with at least six coats of Epoxylite (INS) baked for 10 minutes each at 150°C; bake the last coat for at least 30 minutes.

4. Make a heater wire spiral similar in shape to the coolant tube but with a greater number of loops. Use 0.003 inch diameter insulated silver wire and make the spirals of the wire approximately 0.5 mm apart. Wind the spiral on the sticky side of a piece of masking tape.

5. Cement the heater spiral (HW) to the Epoxylite insulation on the cryoplate. Cover the heater spiral with an additional coat of Epoxylite baked at 150°C for 30 minutes.

6. Attach the thermocouples to the cooling and heating surfaces (TC). (See step 6 under *Cryoprobe* instructions.)

7. Clip the heater wires 2 to 3 mm from the point where they leave the coolant tube. Scrape the insulation from these small wires and solder them to larger diameter 22-gauge heater wires which can carry 4 amperes of current without melting. Remove the excess flux with alcohol and encapsulate the solder joints in dental cement, attaching the dental cement to the coolant tube for rigidity (DC).

8. If metal male Luer-Lok connectors are used to attach the coolant hoses to the cryoplate, drill small holes in their metal surfaces so that the dental cement will more firmly adhere to them at implantation.

9. Solder the metal Luer-Lok connectors (L) to the metal coolant tubes, using a heat sink to avoid melting the dental cement already attached.

10. Cut the lead wires (TC, H) to their proper length and attach contacts.

Materials: stainless steel hypodermic-needle tubing, untempered, 26-gauge, 0.018 inch outside diameter × 0.00425 inch wall thickness; Eutecrod No. 157 silver solder and flux; Epoxylite; silver wire, 0.003 inch diameter, insulated with a 0.001 inch coat; copper wire, 0.005 inch diameter, uninsulated; constantan wire, 0.005 inch diameter, uninsulated; stranded hook-up wire, 22-gauge diameter, insulated; Luer-Lok connectors; contacts; dental cement.

MECHANISMS FOR CRYOGENIC SYSTEM (FIG. 5–12)

Comparator Relay. 1. Use a pico-power comparator relay which has a sensitivity of approximately 15 to 20 microvolts. This sensitivity is required to detect a temperature change of approximately 0.5°C.

2. Attach a power relay across the contacts of the comparator relay. The coil of the relay must be arc-suppressed by placing a diode in reverse polarity across the coil. The contacts of the power relay must be able to carry at least 4 amperes of current.

3. Use an automobile storage battery for the heater current if electrophysiological recording is going to occur during operation of the system.

4. Attach the shaft thermocouple leads to the input of the amplifier (TC 3, 4).

5. Before the cooling ever begins, the balance control of the comparator relay must be adjusted so that its relay is just barely turned off. This setting of the comparator relay ensures that the heater current will be turned off if the shaft temperature rises 0.5°C above the animal's body temperature. Various cut-off levels can be set by the balance control, but they should never allow the shaft temperature to rise more than a few degrees above the animal's body temperature or permanent damage will ensue.

DC Amplifier. 1. Use an operational amplifier which has a 15 to 20 microvolt sensitivity so that resolution of the thermocouple-monitored temperatures is 0.5°C (each thermocouple produces 40 microvolts per degree centigrade difference in temperature between the tip and reference junctions).

2. Attach either the tip or the shaft thermocouple to the input of the recording amplifier. Both the tip and the shaft temperatures can be monitored by switching the input back and forth between the two thermocouples (TC 1, 2, 3, 4).

3. Draw a temperature scale on a thin piece of paper and cement it to the face of the meter (M).

4. When the input to the amplifier is shorted (i.e., zero-voltage input), the balance control (R_3) must be adjusted until the meter reads the known reference-junction temperature. Then the amplifier input is switched to the probe thermocouple, which is submerged in a known temperature bath (e.g., ice water), and the gain control (R_2) is adjusted until the meter reads that known temperature. The value of R_2 should be approximately 50 times R_1. The balance and gain resistors are adjusted several more times until the meter reads the known temperatures exactly.

Heater Control. 1. Use a power transistor which has a 10-ampere current-carrying capacity.

2. Put all the load resistance in the emitter circuit as shown so that the transistor will self-bias and can be turned to the lower current values by the heater-control potentiometer (R_1).

3. In order to monitor the current going through the cryoprobe without having to expend energy carrying the current to and away from

a meter that is remote from the heater power supply, a high resistance meter (M) can be placed across a low resistance shunt (R_2) in the high current pathway, as shown. The meter can then be mounted far away from the shunt and its scale calibrated in units of amperes.

Coolant Delivery. 1. Alcohol is circulated through the cryogenic system by placing it in a steel reservoir tank and forcing it out under pressure.

2. The flow rate of the alcohol through the system is determined by the pressure inside the tank, which can be controlled by a variable pressure regulator. The flow rate can also be controlled by a needle valve placed in the alcohol pathway, but finer control can be achieved by regulating the pressure inside the reservoir tank.

3. The alcohol is circulated through a dry ice and alcohol heat-exchange bath in order to cool the alcohol before it enters the devices implanted in the animal. The heat-exchange bath should be placed as close as possible to the preparation so that the alcohol will not rewarm while passing through the thin polyethylene tubes that carry the coolant to the implanted cryogenic devices.

4. A hypodermic needle can be attached to the coolant hose by wrapping a piece of 28-gauge stainless steel suture wire firmly around the polyethylene tube when it is inserted over the needle (f_W); this fitting will withstand high pressures. The hypodermic needle can then be attached to a Leur-Lok fitting (f_L) soldered onto the stainless steel tubing as it emerges from the heat-exchange bath.

Materials. A, One comparator relay (e.g., PPC-1, Data Device Corporation); 22.5-volt battery to power the above; 50 K ohm, 10-turn potentiometer; power relay with 4-ampere contacts; battery to power the above; arc-suppression diode for above relay coil; 6- or 12-volt automobile storage battery. B, Rotary switch, 2P2T, shorting type, silver contacts; DC operational amplifier (e.g., D-16, Data Device Corporation); two 15-volt batteries to power the above; 50 K ohm, 25-turn potentiometer; 100 K ohm, 25-turn potentiometer; 1 K ohm resistor; 50 microammeter, 2000 ohm. C, Potentiometer, 1500 ohm, 2 watt, linear taper; 10-ampere power transistor, NPN type; 50 microammeter, 2000 ohm. D, 2000 psi steel pressure tank (well cleaned or equipped with 15-micron filter); pressure regulator, variable from 0 to 500 psi; electric valve for on-off control of alcohol flow; 10 feet of stainless steel tubing, 0.125 inch diameter, untempered; Luer-Lok connector, male, metal; polyethylene beaker for dry ice and alcohol bath; 20 feet of Adams Intramedic polyethylene tubing, PE100.

SIMPLE CRYOGENIC SYSTEM (FIG. 5–13)

1. The pressurized coolant is supplied to the system by two syringes. A small 5-cc syringe is used to circulate the pressurized coolant, and when this syringe is empty it can be refilled quickly through a stopcock (S) by a larger 50-cc syringe.

2. The alcohol is forced through a piece of polyethylene tubing (P_1, P_2, P_3) exactly 3 meters long (use only Adams Intramedic tubing,

PE100), through a cryogenic device (C) made from 4 inches of 26-gauge stainless steel tubing, and then through a coolant-return polyethylene tube (P_4) which is 1 m long. The middle third of the tubing (P_2) is looped into a coil consuming exactly 1 m and is submerged in a dry ice and alcohol bath. The lengths of tubing, P_1, P_2, P_3 and P_4, are all exactly 1 m.

3. The cryogenic device (C) can be formed into a cryoplate, as shown, or into a subcortical cryoprobe whose tip and shaft are both cooled. The same amount of the 4 inches of tubing must be in contact with the tissue for both devices. One inch of tubing extends above the surface of the skull for attaching the coolant hoses. For a cryoplate, 3 inches of tubing touch the tissue, but only on one side. For a cryoprobe, 1.5 inches is implanted subcortically, touching the tissue on both sides. A bilateral set of probes is required for use in the rat brain.

4. The alcohol flows out of the system into a graduated cylinder in order that the flow rate of the coolant circulating through the system can be measured. The temperature of the surface of the cryogenic device (C) is proportional to the flow rate, and the calibrated temperature values for this standardized system are given in Table 5-2.

5. Since the flow rate cannot be measured instantaneously, the temperature is difficult to control with finger pressure on the syringe. The instantaneous flow rate can be determined by using a simple manometer (M) to measure the pressure exerted by the syringe. The syringe pressure calibrated in terms of temperature of the implanted cryoplate surface is also given in Table 5-2. To make the manometer, drill a small hole through the side of a metal female Luer-Lok connector (F_L) and solder into place two 0.75-inch lengths of 22-gauge, stainless steel tubing, as shown. Attach the coolant hose to one of these 22-gauge tubes and a 15-inch segment of polyethylene tubing to the other. Use only PE100 tubing. Attach the tubing with a wire fitting made by wrapping 28-gauge, stainless steel suture wire over the polyethylene tubing while

TABLE 5–2 CALIBRATIONS FOR STANDARD CRYOGENIC SYSTEM*

T	FR	Mm	Mm	Pm	Me	Me	Pe
°C	ml/min	mm	in.	lb/in.²	mm	in.	lb/in.²
+25	1.6	51	2.0	5	99	3.9	10
+20	2.6	86	3.4	8	147	5.8	17
+15	3.6	132	5.2	14	185	7.3	24
+10	4.6	168	6.6	20	214	8.4	32
+ 5	5.6	196	7.7	26	231	9.1	42
0	6.6	218	8.6	34	246	9.7	56
− 5	7.6	231	9.1	42	256	10.1	70
−10	8.6	242	9.5	52	266	10.5	96
−15	9.6	249	9.8	61	−	−	−
−20	10.6	259	10.2	74	−	−	−

*To convert flow rate (FR) or pressure to temperature (T) on the surface of the implanted cryogenic device. The pressure (Pm) for the methanol coolant is proportional to the height of the meniscus (Mm) in the manometer, and similarly the pressure (Pe) for the ethanol coolant is proportional to the height of its meniscus (Me). The level of the meniscus is the height of the liquid displacing a 12-inch or 305-mm air column.

it is inserted over the metal cannula (F_W). The tubing may first have to be heated and pulled to narrow the inside diameter, or a piece of PE50 tubing can be inserted into the PE100 before it is wrapped. Draw a 12-inch column of air into the 15-inch tube and then clamp it off so that it will not leak. Use a hemostat clamp if none other is available. The length of the 12-inch column of air is proportional to the pressure, the flow rate and the temperature of the surface of the cryoplate; it is calibrated for this standard system in terms of temperature as given in Table 5-2. Be careful not to create or release the pressure in the manometer too rapidly or bubbles will form in the air column; this column must be purged and then redrawn and reclamped if bubbles do form.

Materials: 5-cc syringe; 50-cc syringe; 2-way stopcock with Luer-Lok fittings; Luer-Lok adaptor, female, metal; 3 inches of stainless steel hypodermic-needle tubing, tempered, 22-gauge, 0.028 inch outside diameter × 0.006 inch wall thickness; 4 inches of stainless steel tubing, untempered, 26-gauge, 0.018 inch outside diameter × 0.00425 inch wall thickness; 12 feet of Adams Intramedic polyethylene tubing, PE100; Eutecrod No. 157 silver solder and flux; 3 feet of 28-gauge, stainless steel suture wire; clamp (hemostat); 12-inch ruler; 10-cc graduated cylinder; 250-cc beaker; 1 liter ethyl or methyl alcohol; 10 lbs. dry ice.

CHEMOPROBE (FIG. 5–14)

1. To make the outer cannula (O), cut a piece of 26-gauge, tempered, stainless steel tubing to the desired length. This length is 2 mm shorter than the chosen implant length plus 8 mm to extend above the surface of the skull into the connector strip (CS).

2. Using dental cement, attach the outer cannula to a piece of connector strip (CS) cut to be 7 mm tall. Allow the tube to extend 1 mm above the top of the connector strip, as illustrated.

3. To make the inner cannula (I), cut a piece of 30-gauge, tempered, stainless steel tubing slightly longer than the implant depth plus 14 mm to extend above the surface of the skull after implantation into the cannula holder.

4. Using dental cement, attach the inner cannula to a piece of connector strip 6 mm tall. Do not allow the dental cement (DC) to run all the way down into the hole in the strip because the top part of the outer cannula tube must be able to fit up inside it approximately 1 mm.

5. Using a No. 54 drill, enlarge the two lateral holes in the connector strip attached to the inner cannula. Using a No. 56 drill and a 0-80 tap, thread the two lateral holes in the connector strip attached to the outer cannula.

6. Insert the inner cannula in the outer one and then screw in two 0-80 × 0.5-inch fillister-head machine screws (M), as illustrated.

7. Make several shims (S) which can be inserted and removed without having to remove the screws. These shims should be made out of metal which is approximately 0.5 mm thick.

8. Screw the inner and outer cannulas together firmly with the

shims in place, and then cut off the inner cannula to the desired implant length. Ream out and smooth the edges. To clean out the inner cannula, squirt alcohol through it with a syringe attached to a polyethylene tube which has a wire fitting (see Fig. 5-13, F_W) appropriate in size to connect to the small, 30-gauge tubing.

9. To load the inner cannula for chemostimulation, first place sterilized crystals of the chemical-stimulating compound between two sterilized microscope slides and, pressing with your fingers, move the glass slides back and forth to grind the crystals into a fine powder. Then tap the sterilized inner cannula repeatedly into a thin layer of the ground chemical substance spread out on one of the slides.

Materials: stainless steel tubing, tempered, 30-gauge, 0.006 inch outside diameter × 0.002 inch wall thickness; stainless steel tubing, tempered, 26-gauge, 0.018 inch outside diameter × 0.00425 inch wall thickness; dental cement; machine screws, fillister head, 0-80 × 0.5 inch; connector strips, 6 mm height and 12 mm height; metal shim material, 0.5 mm thick.

LIST OF DISTRIBUTORS FOR MATERIALS

Johnson, Matthey, and Co., Inc.
608 Fifth Ave.
New York, New York 10020

Stainless steel wire, type No. 304, nontempered, Diamel insulated, 0.0048″ diameter and 0.0092″ diameter; tungsten wire, 0.015″ diameter; silver wire, 0.003″ diameter, Diamel insulation, 0.001″ thick

Medwire Corporation
121 S. Columbus Ave.
Mt. Vernon, New York 10553

Platinum wire, 90% platinum, 10% iridium, wire diameter 0.007″

Omega Engineering Co.
Box 4047 Springdale Station
Stamford, Connecticut 06907

Copper wire, uninsulated, 0.005″ diameter; constantan wire, uninsulated, 0.005″ diameter

V. Mueller and Co.
6818 Del Rio
Houston, Texas 77021

Stainless steel suture wire, 28 ga., nontempered

Tubesales and Co.
Arlington Bank and Trust Bldg.
Room 309
Arlington, Texas 76010

Stainless steel tubing, hypodermic-needle tubing, No. 304; for tempered, specify full hardened; for nontempered, specify hypoflex,
26 ga.: 0.018″ × 0.00425″
22 ga.: 0.028″ × 0.006″
30 ga.: 0.006″ × 0.002″

Hamilton Electronic Sales
1216 W. Clay
Houston, Texas 77019

Contacts, male, Amphenol No. 220-PO2; connector strips, 6 mm No. 221-1260, 12 mm No. 221-1160

Epoxylite Corporation
Box 3387
South El Monte, California 91733

Epoxylite, No. 6001 (inert), No. 6001M (modified)

Eutectic Welding and Alloys Co.
5917 Armour Dr.
Houston, Texas 77020

Eutecrod No. 157 silver solder, $1/16''$;
Eutectic No. 157 flux

R. P. Gallien and Son
220 West 5th St.
Los Angeles, California 90013

Machine screws, stainless steel, fillister head:
No. 2-56 $\times$ $3/16''$
No. 0–80 $\times$ $3/16''$
No. 0–80 $\times$ $1/2''$

The William Getz Corporation
7512 S. Greenwood Ave.
Chicago, Illinois 60619

Dental cement, Getz Tru-Cure "Self-Cure" denture-base material, pink; liquid for above, fast drying

Mills Hospital Supply Co.
2918 West Dallas
Houston, Texas 77019

Adams Intramedic polyethylene tubing:
PE100
PE50

REFERENCES

Carpenter, M. B., and Whittier, J. R. (1952). Study of methods for producing experimental lesions of the central nervous system with special reference to stereotaxic technique. *J. Comp. Neurol.* 97:73-132.

Dondey, M., Albe-Fessard, D., and Le Beau, J. (1962). Premières applications neurophysiologiques d'une méthode permettant le blocage électif et réversible de structures centrales par réfrigération localisée. *Electroenceph. Clin. Neurophysiol.* 14:758–763.

Skinner, J. E., and Lindsley, D. B. (1967). Electrophysiological and behavioral effects of blockade of the nonspecific thalamo-cortical system. *Brain Res.* 6:95-118.

Skinner, J. E., and Lindsley, D. B. (1968). Reversible cryogenic blockade of neural function in the brain of unrestrained animals. *Science 161*:595-597.

CHAPTER 6

STEREOTAXIC ATLAS
OF THE RAT BRAIN

Several atlases of the rat brain have been published within the last few years, each making a scientific contribution not contained in the others. The present atlas brings together many of these contributions under one cover, and in addition adds a few improvements: (1) the use of a fixative which does not shrink and distort the brain tissue used to make the atlas, (2) the extension of the atlas planes posteriorly into the medulla, (3) two sets of stereotaxic reference points, and (4) statistical computations on the various anatomic measurements relevant to the construction and use of a stereotaxic atlas.

SUBJECTS AND METHODS

Twelve rats were selected at random from the young adult population of the albino, Longevans strain used by the University of California and bred and sold by the Simonson Company in Gilroy, California. These subjects were chosen to range in age from 120 to 150 days, which was considered to be the minimum age of the young adult category in which the brain size seems to remain stable. Half of the subjects were male and half female. The average weight and standard deviation of the males was 493.67 ±43.4 grams, and the average for the females was 340.83 ±7.4 grams.

The subjects were anesthetized and placed in a stereotaxic instrument, using ear bars whose puncture-proof tips were beveled 62 degrees from the long axis. The midline was determined by the symmetrical adjustment of the ear bars in the stereotaxic instrument after the animal was mounted. The top of the incisor bar was 5 mm above the center of the ear bars. Three marker pins were placed in the brain: two in the coronal planes at bregma and at the center of the tip of the ear bars,

195

and the other in the horizontal zero plane 5 mm above the tip of the ear bars. The coronal pins were 2 mm lateral and the horizontal pin 5 mm lateral, and all were on the same side of the brain. After the pins were in place, the animals were injected with one of three fixatives via the heart (40 per cent formaldehyde, 4 per cent formaldehyde, or 2 per cent gluteraldehyde plus 2 per cent formaldehyde), and all injections resulted in good brain perfusion. After fixation, the brains were blocked in the coronal plane with a razor saw while the animal was still in the stereotaxic instrument, and then each brain was immediately extracted from its brain case.

The three perfusion groups were matched by age, weight and sex. The brains of the group receiving the 40 per cent formaldehyde perfusate were immediately frozen after extraction and cut with a freezing microtome in the coronal planes. The other brains were left to soak in their perfusate for 24 hours before cutting with the freezing microtome.

It was found that the brains injected with 40 per cent formaldehyde and then cut right away were neither shrunken nor distorted. The cortex contained the outlines of adjacent sinus vessels, arteries and fissure and bone indentations which were absent or reduced in the tissue of other fixative groups. By comparing the distance between the tips of the pins implanted in each brain in the coronal planes, at the bregma and the intra-aural line, and the distance between the outlines of their tips, which were left in the tissue and observed after sectioning, it was determined that the 40 per cent formaldehyde perfusate produced only 6 per cent shrinkage, while the routine histological method of perfusing and fixing the brain with 4 per cent formaldehyde resulted in as much as 30 per cent shrinkage. The 2 per cent gluteraldehyde and 2 per cent formaldehyde mixed perfusate used by electron microscopists to minimize osmotic distortion of membranes resulted in 22 per cent shrinkage. A sample of the tissue perfused by 40 per cent formaldehyde was embedded in araldite and sectioned for observation under the electron microscope. This tissue did not show any abnormal distortion of the membranes but did show an observable extracellular space which is often absent in tissue perfused with the 2 per cent gluteraldehyde and 2 per cent formaldehyde mixture.

All the brains were cut in serial order in 100 micron sections, and each section was mounted on a slide by floating it in a bowl filled with its particular perfusate. The section was then placed in a photographic enlarger and printed as if it were a photographic negative (Guzman-Flores et al., 1958). Each tissue section was printed at a carefully calibrated enlargement of 10 times. The section was then stained with thionine, and the next section cut in the serial order was stained with a Weil myelinated fiber stain. Sections were photographed and stained in this manner throughout the entire extent of the brain. The atlas plates show, at 0.5 mm intervals, the photographic enlarger prints of the representative brain chosen for the atlas, and the associated line drawings show the location of the nuclear groups observed after the tissue was stained. The relative position of the nuclear groups had to be reconstructed on the line drawings because the stained tissue was shrunken and distorted by the staining procedure.

NOMENCLATURE

If the nuclear groups could be seen clearly, they were outlined on the drawing with *dashed lines*, but if they were ambiguous and not clearly defined they were outlined with *dotted lines*. The myelinated fiber tracts and gross anatomical structures were outlined with *solid lines*. Note that the abbreviations of the nuclear groups are all in *lower case letters*, and fiber tracts and gross anatomical structures are labeled with *upper case letters*. All the brain structures in this atlas have been named using the conventional nomenclature specified by the international congress on anatomic nomenclature (Excerpta Medica Foundation, 1966). Those few brain structures which have not as yet been named by the international congress have been labeled with an *asterisk* and named according to common usage. The published work of Wunscher et al. (1965) was particularly useful in identifying the ambiguous and unnamed structures of the reticular formation, and their nomenclature was followed in labeling this particular part of the brain. Other works were consulted to help identify the other structures (Cajal, 1896; Gurdjian, 1927; Krieg, 1954; DeGroot, 1959; Massopust, 1961; Craigie, see Zeman and Innes, 1963; Albe-Fessard et al., 1966; Yoshikawa, 1968).

The cytoarchitectonic areas of the cortex have been studied in the rat brain by only a few investigators. Although there is fair agreement among them, it is exceedingly difficult for unskilled persons to create similar cytoarchitectonic field maps. The larger boundary areas on the cortex seen in the figure and outlined by solid lines are in fair agreement between Krieg (1954), Kuhlenbeck et al. (1960), and ourselves; however, the smaller regions show a great deal of variability between animals in our material, and are not unambiguously defined. The Brodman areas (1909) numbered in this figure were adapted from Krieg's work to our own stereotaxic brain atlas, using Brodman's criteria on our own material. It must be clear that we were not always certain that we could see or agree upon all these boundaries separating the cytoarchitectonic areas described, but in order to provide a convenient system for tabulating the cortex, we have acquiesced to higher authority (Table 6-1).

STEREOTAXIC INSTRUMENT REFERENCE POINTS

The horizontal stereotaxic plane is determined by the location of the *top* of the incisor bar and the *center* of the ear bars. The incisor bar is located 5 mm above the center of the ear bars on the horizontal zero plane. The coronal and lateral planes are at right angles to the horizontal plane and to each other. The zero lateral plane is the plane midway between the tips of the ear bars which are centered symmetrically in the

TABLE 6–1   SURFACE STRUCTURES

Cerebral Cortex
Cingulate (23, 24, 25, 25a, 29b, 29c, 32)
Frontal–sensorimotor (4, 6, 8, 8a, 10, 11)
Pyriform (51a, 51b, 51e, 51f, 51g, 51h, Tol)
Retrohippocampal (27, 28a,* 35, 49,* Amm, Fd, Sub)
Parietal (1, 2, 2a, 3, 7, 39, 40)
Occipital (17, 18, 18a, 36)
Insular (13, 14)
Temporal (20, 41)

Cerebellar Cortex and Fissures
Cerebellar Cortex
  Lingula (L)
  Centralis (Ce)
  Culmen (Cu)
  Simplex (S)
  Declive (D)
  Ansiformis (ANS)
  Paramedianus (PM)
  Tuber (T)
  Pyramis (P)
  Uvula (U)
  Nodulus (N)
  Paraflocculus (PF)

Cerebellar Fissures
  Fissura prima (FP)
  Fissura prepyramidalis (FPP)
  Fissura posterolateralis (FPL)

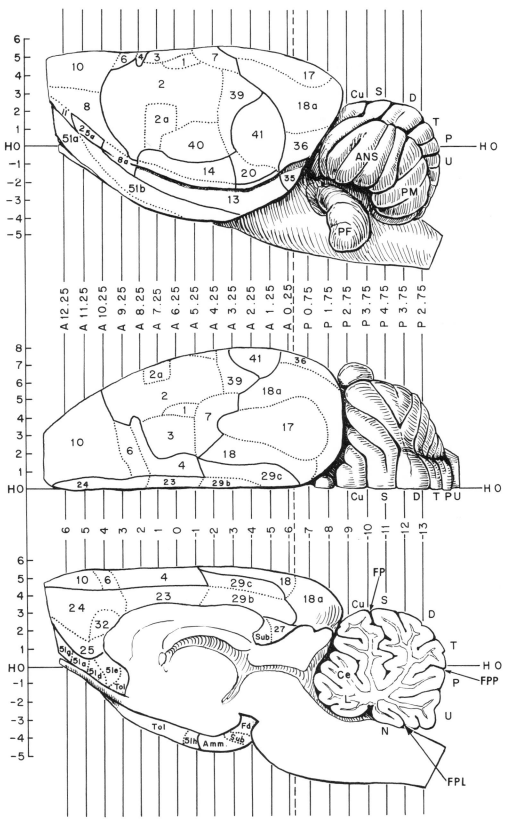

Figure 6–1

stereotaxic instrument after the animal is mounted. The zero coronal plane runs through the center of the tips of the ear bars. The origin of the three-dimensional coordinate system is therefore located in the approximate center of the brain, and the anterior and posterior commissures are in the same horizontal plane. This particular orientation of the stereotaxic planes results in coronal sections comparable to those in brain atlases of other mammals. Since the origin is in the center of the brain, the implant coordinates can be corrected for brain size by multiplying the atlas coordinates by a single linear correction factor.

BREGMA-MIDLINE-CORTICAL SURFACE REFERENCE POINTS

An alternative set of atlas coordinates which has been constructed has two advantages: less variability of the reference points with respect to brain structures, and less effect on actual placement of electrode tips in improperly mounted animals. The variability of bregma with respect to the brain was determined to be less than that of the auditory meatus. The distance between the pin track made in the tissue at bregma and the nearby anterior commissure, and the distance between the pin track made by the pin implanted in the zero coronal plane and the adjacent posterior commissure, were measured for each brain and the standard deviations were computed. The standard deviation of bregma with respect to the brain was 0.3 mm and that for the zero horizontal plane or auditory meatus was 0.4 mm. The variability of the distance between the auditory meatus and bregma was 0.5 mm, which indicates, according to statistical theory, that these cranial reference points were varying independently from each other (for independent variabilities: $SD_A^2 + SD_B^2 = SD_C^2$; $0.3^2 + 0.4^2 = 0.5^2$).

The lateral distance of the coronal pin tracks in the tissue from the midline of bilateral symmetry had the greatest variability (SD = 0.6 mm), which indicates that the symmetrical centering of the ear bars after the animal is mounted in the stereotaxic instrument does not define very accurately the origin of the lateral plane. In a group of six additional animals in which the electrodes were located laterally by measuring from the midline cranial sutures, it was determined that the standard deviation of the distance between the pin tracks in the tissue and the midline of bilateral symmetry was only 0.4 mm. It is therefore recommended that the lateral coordinates always be measured from the center of the midline cranial sutures.

The depth of implantation is measured from the top of the exposed cortex when the bregma and midline suture points are used for coordinate references.

If the animal is improperly mounted in the stereotaxic ear bars, the anterior-posterior, lateral and depth coordinates will not be so much in error as those others which are referred to the stereotaxic ear-bar tips. When using the bregma-midline-cortical surface reference points, only

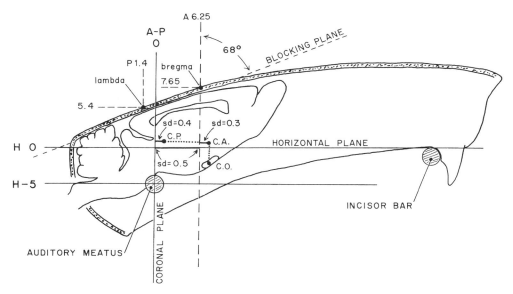

Figure 6–2

the angle of orientation of the coronal planes will be affected, and often this will not alter significantly the placement of the electrodes.

It was found, using statistical tests, that there were no significant correlations between age, sex or body weight and the size of the brain or the locations of the bregma and lambda points on the skull. Of course, older animals weighed more than younger ones, and males weighed more than females, but there were no differences in their brain sizes.

BLOCKING PLANE

After a brain has been perfused and extracted from its brain case, it can be laid upside down on a horizontal surface resting on the cerebellum and cerebrum. The coronal stereotaxic plane will then be at an angle of approximately 68 degrees to the horizontal surface in the posterior direction. One can block the brain at this angle for sectioning in the coronal planes after the brain has been extracted, which is much more convenient than using a razor saw to block the brain while it is still in the stereotaxic instrument.

SUMMARY

SELECTION OF SUBJECTS

The animals should all be between 120 and 150 days of age and weigh 340 ±10 grams if female and 490 ±40 grams if male. They should be chosen from a population of the Longevans variety of laboratory rats, and preferably from the Simonson Company albino strain. Older rats can be used, but they will weigh more, have tougher skins, greater periosteal tissue, larger temporal muscle insertions and thicker skulls — none of which are desirable. Other strains of rats can be used, but coordinate correction factors may need to be determined.

REFERENCE POINTS

The top of the incisor bar should be 5 mm above the center of the ear-bar tips. Puncture-proof ear bars whose tips have a bevel of 62 degrees from their long axis should be used. If the stereotaxic instrument coordinate system is used, the animal *must be properly mounted* in the stereotaxic instrument. After the animal is mounted and the scalp incision made, check to see if bregma is within 6.25 ±0.5 mm anterior to the zero coronal plane and 7.65 ±0.5 mm above the horizontal zero plane. Lambda is 1.4 ±0.5 mm posterior and 5.4 ±0.5 mm above the respective coronal and horizontal zero planes. If the bregma-midline-cortical surface reference points are used, it is less important that the animal be properly mounted; however, lambda should be within 2.25 ±0.5 mm lower than bregma in the horizontal planes.

ANTERIOR-POSTERIOR COORDINATES

Each plate of the atlas represents a coronal plane in 0.5 mm steps and has two sets of numbers in the upper right corner. The uppermost of these are marked + and −, which mean, respectively, anterior and posterior to bregma. The lower set of numbers are marked A and P, which mean, respectively, anterior and posterior to the auditory meatus.

A single brain was used to make the atlas. The size of this brain was exactly the mean of all the subjects. It was perfused with 40 per cent formaldehyde, and then quickly frozen and sectioned. Each section was photographed by placing it in a photographic enlarger and printing it at an enlargement of 10 times as if it were a photographic negative. It was determined that the 40 per cent formaldehyde perfusate shrank the tissue by 6 per cent, so the scale on which the photograph appears has been adjusted by 6 per cent, as have the anterior-posterior coordinates.

NOMENCLATURE

The names of each structure labeled in the atlas are according to the international convention of nomenclature. Some of the more common names also have been included, and are shown in parentheses. The nuclear structures are labeled in lower case letters, and the gross structures and fiber tracts are labeled in upper case letters. Structures which have not been named by the international committee on nomenclature are named according to common usage and labeled with asterisks. The easily identified and well differentiated nuclear structures are indicated on the line drawings by dashed lines, and the hard to identify nuclei are outlined by dotted lines. The fiber tracts are outlined by solid lines.

REFERENCES

Albe-Fessard, D., Stutinsky, F., and Libouban, S. (1966). *Atlas stereotaxique du diencephale du rat blanc.* Paris, Editions du Centre National de la Recherche Scientifique.

Brodman, K. (1909). *Vergleichende Lokalisationslehre der Grosshirnrinde.* Leipzig, Johann Ambrosius Barth.

Cajal, S. Ramon y. (1896). *Beitrag zum Studium der Medulla Oblongata.* Leipzig, Johann Ambrosius Barth.

DeGroot, J. (1959). *The Rat Forebrain in Stereotaxic Coordinates.* Amsterdam, N. V. Noord-Hollandsche Uitgevers Maatschappij.

Excerpta Medica Foundation (1966). *Nomina Anatomica.* Amsterdam, Mouton and Co.

Gurdjian, E. S. (1927). The diencephalon of the albino rat. *J. Comp. Neurol. 43:*1-114.

Guzman-Flores, C. M., Alcaraz, M., and Fernandez-Guardiola, A. (1958). Rapid procedure to localize electrodes in experimental neurophysiology. *Bol. Inst. Estud. Med. Biol.* (Mexico) *16:*26-31.

Krieg, W. J. S. (1954). *Collected Papers Relating to the Cerebrum.* Springfield, Ill., Charles C Thomas.

Kuhlenbeck, H., Szekely, E. G., and Spuler, H. (1960). Some remarks on the zonal patterns of mammalian cortex cerebri as manifested in the rabbit: Its relationship with certain electrocorticographic findings. *Confin. Neurol. 20:*407-423.

Massopust, L. C., Jr. (1961). Diencephalon of the rat. *In* Sheer, D. E. (ed.): *Electrical Stimulation of the Brain.* Austin, University of Texas Press, pp. 182-202.

Wunscher, W., Schober, W. and Werner, L. (1965). *Architektonischer Atlas vom Hirnstamm der Ratte.* Leipzig, S. Hirzel.

Yoshikawa, T. (1968). *Atlas of the Brains of Domestic Animals.* University Park, Pennsylvania State University Press.

Zeman, W., and Innes, J. R. M. (1963). *Craigie's Neuroanatomy of the Rat.* New York, Academic Press, Inc.

LIST OF ABBREVIATIONS FOR RAT ATLAS
INTERNATIONAL NOMENCLATURE OF BRAIN STRUCTURES

NUCLEI

a	N. accombens
abl	N. amygdaloideus basalis, pars lateralis
abm	N. amygdaloideus basalis, pars medialis
ac	N. amygdaloideus centralis
aco	N. amygdaloideus corticalis
al	N. ansae lenticularis
ala	N. amygdaloideus lateralis, pars anterior
alp	N. amygdaloideus lateralis, pars posterior
am	N. amygdaloideus medialis
arl	N. arcuatus lateralis
arm	N. arcuatus medialis
ca	N. commissurae anterioris
ccgm	N. centralis corporis geniculati medialis
cfd	N. commissurae fornicis dorsalis
cl	N. claustrum
cm	N. centre median
cp	N. caudatus putamen
ctl	N. corporis trapezoidei lateralis
ctm	N. corporis trapezoidei medialis
cu	N. cuneiformis (mesencephalic reticular formation)
d	N. Darkschewitsch
degl	N. dorsalis corporis geniculati lateralis
dr	N. dorsalis raphes
dt	N. dorsalis tegmentalis (von Gudden)
ep	N. entopeduncularis
fm	N. paraventricularis (filiformis), pars magnocellularis
fp	N. paraventricularis (filiformis), pars parvocellularis
g	N. gelatinosus
ha	N. anterior hypothalami
hdd	N. dorsomedialis hypothalami, pars dorsalis
hdv	N. dorsomedialis hypothalami, pars ventralis
hl	N. lateralis hypothalami
hp	N. posterior hypothalami
hpv	N. periventricularis hypothalami
hvm	N. ventromedialis hypothalami
hvma	N. ventromedialis hypothalami, pars anterior
hvmc	N. ventromedialis hypothalami, pars centralis
hvml	N. ventromedialis hypothalami, pars lateralis
hvmm	N. ventromedialis hypothalami, pars medialis
hvmp	N. ventromedialis hypothalami, pars posterior

i	N. interstitialis (Cajal)
icp	N. intracommissuralis commissurae posterioris
idtv	N. interstitialis decussationis tegmenti ventralis
imcp	N. interstitialis magnocellularis commissurae posterioris
ip	N. interpeduncularis
lc	N. linearis, pars caudalis
lh	N. habenulae lateralis
li	N. linearis, pars intermedialis
lr	N. linearis, pars rostralis
mcgm	N. marginalis corporis geniculati medialis
mh	N. medialis habenulae
ml	N. mamillaris lateralis
mml	N. mamillaris medialis, pars lateralis
mmm	N. mamillaris medialis, pars medialis
mp	N. mamillaris posterior
mpcs	N. marginalis pedunculi cerebellaris superioris
mpl	N. mamillaris prelateralis
mr	N. medianus raphes
n III	N. n. oculomotorii
n III ewc	N. Edinger-Westphal n. oculomotorii, pars caudalis
n III ewr	N. Edinger-Westphal n. oculomotorii, pars rostralis
n III pr	N. n. oculomotorii principalis
n IV	N. n. trochlearis
n V m	N. motorius n. trigemini
n V s	N. sensorius n. trigemini
n V spo	N. oralistractus spinalis n. trigemini
n V tm	N. tractus mesencephalici n. trigemini
n VI	N. n. abducentis
n VII	N. n. facialis
n VIII l	N. vestibularis lateralis
n VIII m	N. vestibularis medialis
n VIII s	N. vestibularis superior
oad	N. olfactorius anterior, pars dorsalis
oae	N. olfactorius anterior, pars externa
oal	N. olfactorius anterior, pars lateralis
oam	N. olfactorius anterior, pars medialis
oap	N. olfactorius anterior, pars posterior
ol	N. tractus olfactorii lateralis
ols	N. olivaris superior
p	N. pretectalis
pbl	N. parabrachialis lateralis
pbm	N. parabrachialis medialis
pd	N. premamillaris dorsalis
pf	N. parafascicularis
pl	N. pontis, pars lateralis
pm	N. pontis, pars medialis
pol	N. preopticus lateralis
pom	N. preopticus medialis
poma	N. preopticus magnocellularis
pome	N. preopticus medianus
pop	N. preopticus periventricularis
posc	N. preopticus, pars suprachiasmatica

pp	N. pretectalis profundus
pt	N. paratenialis
pv	N. premamillaris ventralis
pvr	N. periventricularis rotundocellularis
pvs	N. periventricularis stellatocellularis
r	N. ruber
re	N. reuniens
rh	N. rhomboideus
rpc	N. reticularis pontis caudalis
rpo	N. reticularis pontis oralis
rtp	N. reticularis tegmenti pontis
s	N. suprageniculatus
sc	N. suprachiasmaticus
scp	N. subcommissuralis commissurae posterioris
sf	N. septalis fimbrialis
sl	N. septi lateralis
sm	N. septi medialis
so	N. supraopticus
spf	N. subparafascicularis
st	N. interstitialis striae terminalis
sum	N. supramamillaris
sut	N. subthalamicus
tad	N. anterior dorsalis thalami
tam	N. anterior medialis thalami
tav	N. anterior ventralis thalami
td	N. tractus diagonalis (Broca)
tl	N. lateralis thalami
tlp	N. lateralis thalami, pars posterior
tm	N. medialis thalami
tml	N. medialis thalami, pars lateralis
tmm	N. medialis thalami, pars medialis
tob	N. tractus optici basalis (Gillilan)
tol	N. tractus optici, pars lateralis
tom	N. tractus optici, pars medialis
tpm	N. posteromedianus thalami
tpo	N. posterior thalami
tr	N. reticularis thalami
ts	N. triangularis septi
tuo	N. of tuberculum olfactorium
tv	N. ventralis thalami
tvd	N. ventralis thalami, pars dorsomedialis
tvm	N. ventralis medialis thalami, pars magnocellularis
tvp	N. ventralis medialis thalami, pars parvocellularis
vcgl	N. ventralis corporis geniculati lateralis
vcll	N. ventralis caudalis lemnisci lateralis
vrll	N. ventralis rostralis lemnisci lateralis
vt	N. ventralis tegmenti

FIBER TRACTS AND GROSS STRUCTURES

A	Alveus hippocampi
AA	Area amygdala anterior
AC	Aqueductus cerebri (Sylvii)
ACU	Area cuneiformis
AL	Ansa lenticularis
APM	Area pretectalis medialis
AT	Area transitionalis corticoamygdaloidea
AVT	Area ventralis tegmenti (Tsai)
BCI	Brachium colliculi inferioris
BCS	Brachium colliculi superioris
BOA	Bulbus olfactorius accessorius
C	Cingulum
CA	Commissura anterior
CAA	Commissura anterior, pars anterior
CAE	Capsula externa
CAI	Capsula interna
CAIR	Capsula interna, pars retrolenticularis
CAP	Commissura anterior, pars posterior
CC	Crus cerebri
CCI	Commissura colliculorum inferiorum
CCS	Commissura colliculorum superiorum
CE	Cortex entorhinalis
CFD	Commissura fornicis dorsalis (Commissura hippocampi dorsalis)
CFV	Commissura fornicis ventralis (Commissura hippocampi ventralis)
CH	Commissura habenularum
CHLS	Cerebellum hemispherium lobulus simplex
CHPF	Cerebellum hemispherium paraflocculus
CI	Colliculus inferior
CLA	Corpus cerebelli lobus anterior
CO	Chiasma opticum
CP	Commissura posterior
CPF	Cortex piriformis
CSDD	Commissura supraoptica dorsalis, pars dorsalis (Ganser)
CSDV	Commissura supraoptica dorsalis, pars ventralis (Meynert)
CT	Corpus trapezoideum
CTH	Commissurae thalami
CTV	Commissura tuberis ventralis
CVLC	Cerebellum vermis lobulus centralis
DCT	Decussatio corporis trapezoidei
DPCS	Decussatio pedunculorum cerebellarium superiorum
DTD	Decussatio tegmenti dorsalis
DTV	Decussatio tegmenti ventralis
F	Columna fornicis
FC	Fissura chorioidea
FCI	Fissura circularis rhinencephali
FH	Fimbria hippocampi
FIP	Fossa interpeduncularis

FIV	Foramen interventriculare
FL	Fasciculus longitudinalis
FLC	Fissura longitudinalis cerebri
FLD	Fasciculus longitudinalis dorsalis (Schütz)
FLDG	Fasciculus longitudinalis dorsalis (Schütz), pars tegmentalis
FLDT	Fasciculus longitudinalis dorsalis (Schütz), pars tectalis
FLM	Fasciculus longitudinalis medialis
FMA	Forceps major
FMI	Forceps minor
FMP	Fasciculus medialis prosencephali (medial forebrain bundle)
FMT	Fasciculus mamillothalamicus
FMTG	Fasciculus mamillotegmentalis
FO	Fornix
FOP	Fasciculus opticus
FOR	Formatio reticularis (see CU)
FP	Fibrae pyramidales
FPC	Fornix precommissuralis
FPCP	Fornix precommissuralis, pars preoptica et hypothalamica
FPT	Fibrae pontis transversae
FPVH	Fibrae periventriculares hypothalami
FPVT	Fibrae periventriculares thalami
FR	Fasciculus retroflexus
FS	Fornix superior
F IV	Fibrae n. trochlearis
F V	Fibrae n. trigemini
F V TM	Fibrae tractus mesencephalici n. trigemini
F VII	Fibrae n. facialis
F VIII	Fibrae n. statoacustici
GCC	Genu corporis callosi
GD	Gyrus dentatus
GP	Globus pallidus
H	Habenula
HI	Hippocampus
HIA	Hippocampus, pars anterior
H_1	Forel's field H_1
H_2	Forel's field H_2
I	Infundibulum
IC	Insulae Calleja
ICM	Insula Calleja magna
IG	Indusium griseum
LAME	Lamina medullaris externa thalami
LAMI	Lamina medullaris interna thalami
LAT	Lamina terminalis
LCM	Lamina cellularum mitralium bulbi olfactorii
LCMA	Lamina cellularum mitralium bulbi olfactorii accessorii
LG	Lamina glomerulosa bulbi olfactorii
LGA	Lamina glomerulosa bulbi olfactorii accessorii
LGE	Lamina granularis externa bulbi olfactorii
LGEA	Lamina granularis externa bulbi olfactorii accessorii
LGI	Lamina granularis interna bulbi olfactorii
LL	Lemniscus lateralis
LM	Lemniscus medialis
LMIO	Lamina medullaris interna bulbi olfactorii
LMO	Lamina molecularis bulbi olfactorii

LMOA   Lamina molecularis bulbi olfactorii accessorii
LOC   Locus ceruleus
LPEB   Lamina plexiformis externa bulbi olfactorii
LPIB   Lamina plexiformis interna bulbi olfactorii
LSTE   Lemniscus spinalis tractus spinotectalis
LSTH   Lemniscus spinalis tractus spinothalamicus
LT   Lemniscus trigeminalis

MI   Massa intercalata
N IV   Nervus trochlearis
N V M   Nervus trigeminus motorius
N V S   Nervus trigeminus sensorius
N VIII Nervus statoacousticus

OS   Organon subfornicale
OSC   Organon subcommissurale

P   Pons
PCI   Pedunculus cerebellaris inferior
PCM   Pedunculus cerebellaris medius
PCMA   Pedunculus corporis mamillaris
PCS   Pedunculus cerebellaris superior
PF   Polus frontalis
PIN   Corpus pineale
PTI   Pedunculus thalami inferior

R   Raphe
RCC   Radiatio corporis callosi
RI   Recessus infundibuli
RM   Recessus mamillaris
RO   Recessus opticus
RP   Recessus pinealis
RT   Recessus triangularis
RTM   Radiatio thalami intermedia
RTS   Radiatio thalami superior
R III   Radix n. oculomotorii
R IV   Radix n. trochlearis

S   Subiculum
SAM   Stratum album mediale colliculi superioris
SAP   Stratum album profundum colliculi superioris
SCC   Sulcus corporis callosi
SGCD   Substantia grisea centralis, pars dorsalis
SGCL   Substantia grisea centralis, pars lateralis
SGCV   Substantia grisea centralis, pars ventralis
SGM   Stratum griseum mediale colliculi superioris
SGP   Stratum griseum profundum colliculi superioris
SGPV   Substantia grisea periventricularis
SGS   Stratum griseum superficiale colliculi superioris
SH   Sulcus hippocampi
SLL   Stria longitudinalis lateralis (Lancisi)
SLM   Stria longitudinalis medialis (Lancisi)
SM   Stria medullaris thalami
SNC   Substantia nigra, zona compacta
SNL   Substantia nigra, pars lateralis
SNR   Substantia nigra, zona reticulata
SO   Stratum opticum colliculi superioris

SPCC	Splenium corporis callosi
SR	Sulcus rhinalis
ST	Stria terminalis
STC	Stria terminalis, pars commissuralis
STH	Stria terminalis, pars hypothalamica
STI	Stria terminalis, pars infracommissuralis
STM	Stria terminalis, pars ad striam medullarem
STP	Stria terminalis, pars precommissuralis
SUM	Decussatio supramamillaris
SZ	Stratum zonale thalami
T	Tapetum
TCC	Truncus corporis callosi
TCHL	Tractus corticohabenularis lateralis
TCHM	Tractus corticohabenularis medialis
TCL	Tractus corticohypothalamicus lateralis
TCM	Tractus corticohypothalamicus medialis
TCT	Tractus corticotectalis
TD	Tractus diagonalis (Broca)
THC	Tractus hippocampocorticalis
TI	Tractus infundibularis
TIG	Tractus incertotegmentalis
TIT	Tractus incertotectalis
TL	Tela chorioidea ventriculi lateralis
TO	Tractus opticus
TOB	Tractus opticus basalis
TOC	Tractus olfactocorticalis
TOH	Tractus olfactohypothalamicus
TOHL	Tractus olfactohabenularis lateralis
TOHM	Tractus olfactohabenularis medialis
TOI	Tractus olfactorius intermedius
TOL	Tractus olfactorius lateralis
TOLD	Tractus olfactorius lateralis, pars dorsalis
TOLI	Tractus olfactorius lateralis, pars intermedia
TOLV	Tractus olfactorius lateralis, pars ventralis
TOM	Tractus olfactorius medialis
TPT	Tractus pedunculotegmentalis
TRS	Tractus rubrospinalis
TSC	Tractus septocorticalis
TSCV	Tractus spinocerebellaris ventralis
TSH	Tractus septohabenularis
TSHT	Tractus septohypothalamicus
TST	Tractus septotubercularis + Tractus tuberculoseptalis
TSTH	Tractus striohypothalamicus
TTS	Tractus tectospinalis
TULC	Tuberculum olfactorium, pars corticalis, lamina pyramidalis
TULI	Tuberculum olfactorium, pars interna, lamina polymorphica
TULP	Tuberculum olfactorium, pars corticalis, lamina plexiformis
TUO	Tuberculum olfactorium
VL	Ventriculus lateralis
VLP	Ventriculus lateralis, posterior horn
VMS	Velum medullare superius
VO	Ventriculus olfactorius
VSC	Tractus spinocerebellaris ventralis
V III	Ventriculus tertius
ZI	Zona incerta

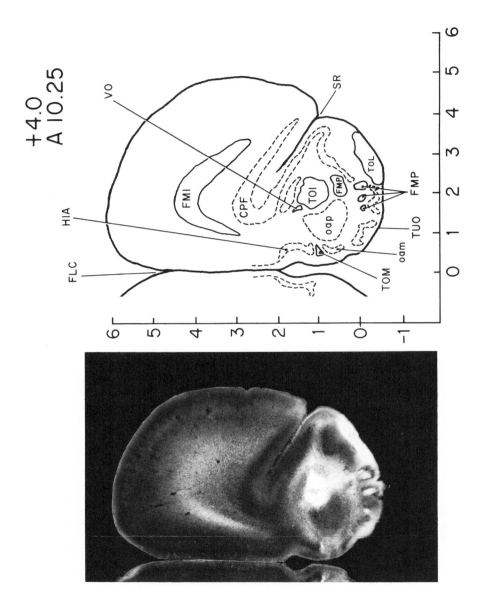

+4.0
A 10.25

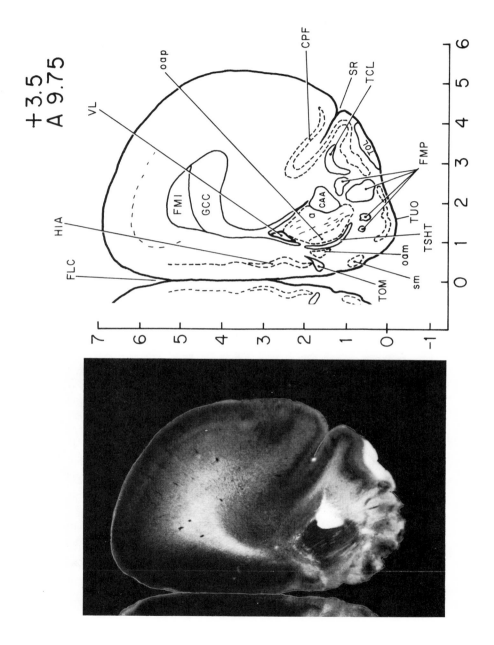

+3.5
A 9.75

213

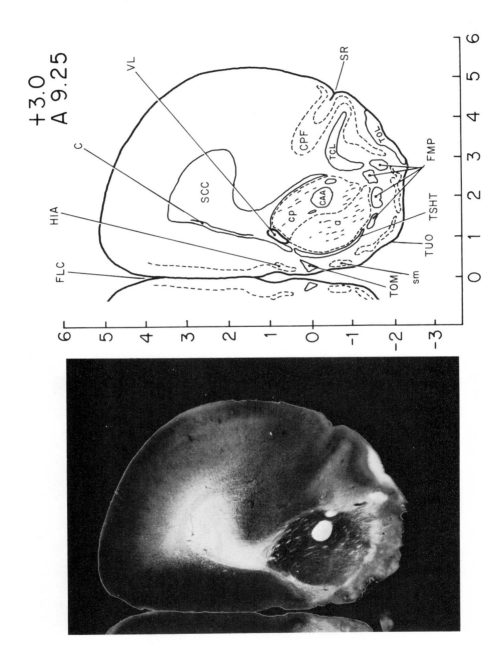

+3.0
A 9.25

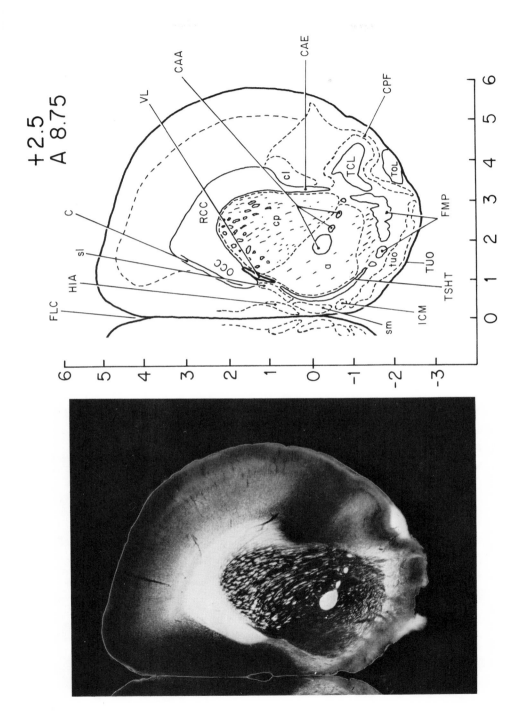

+2.5
A 8.75

215

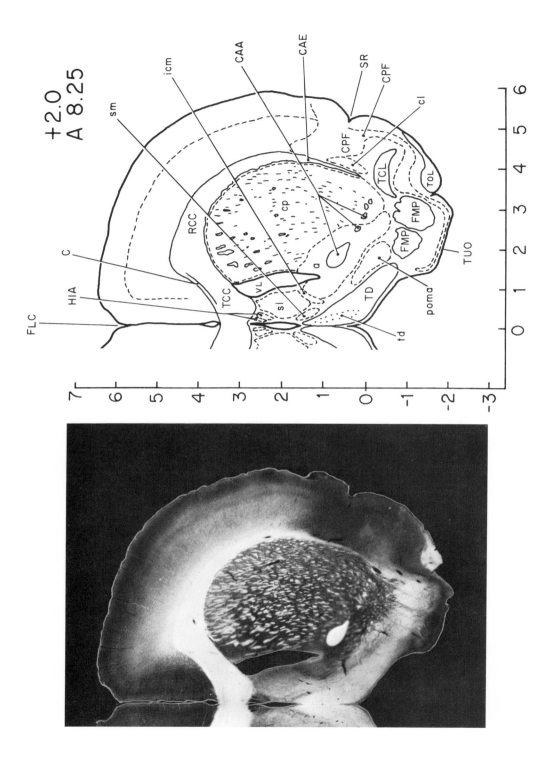

+2.0
A 8.25

216

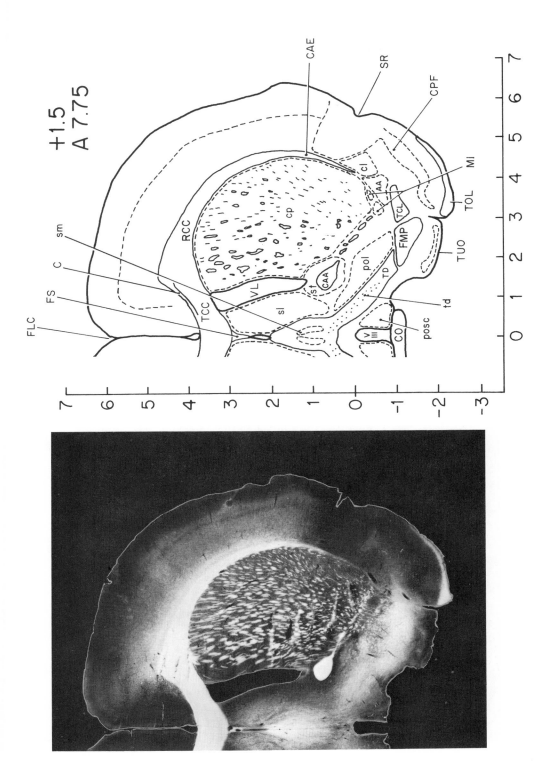

+1.5
A 7.75

217

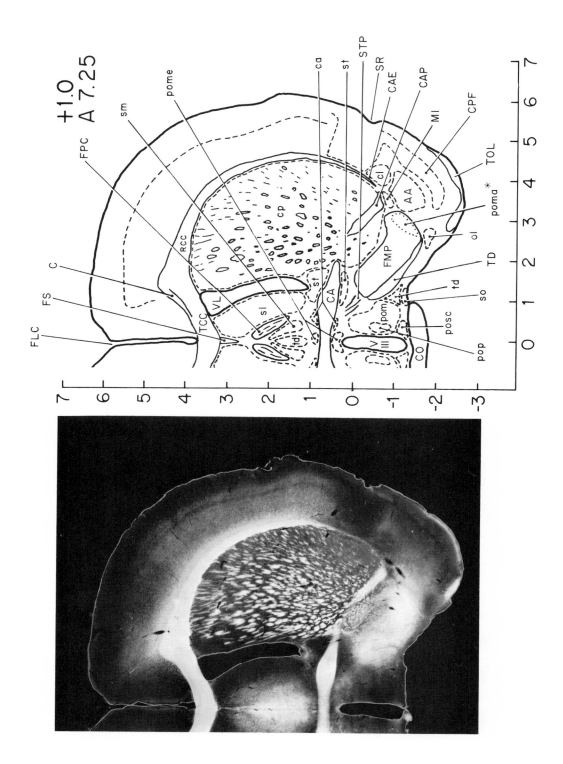

+1.0
A 7.25

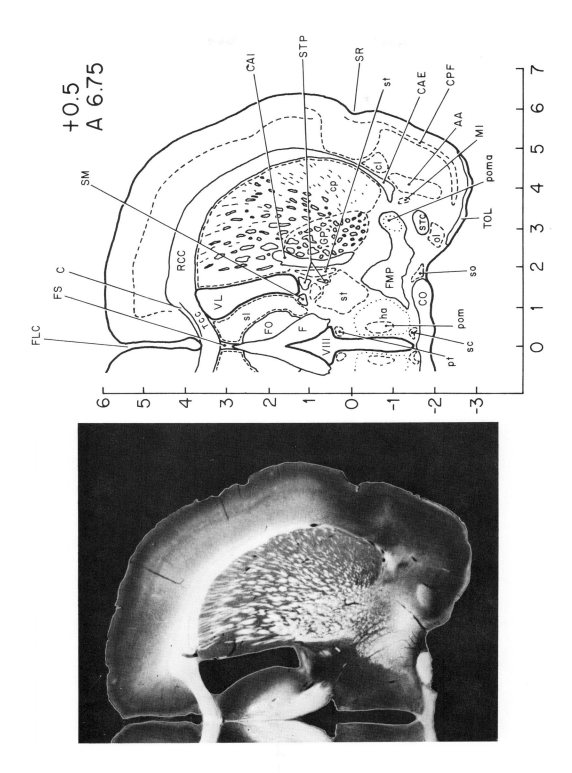

+0.5
A 6.75

219

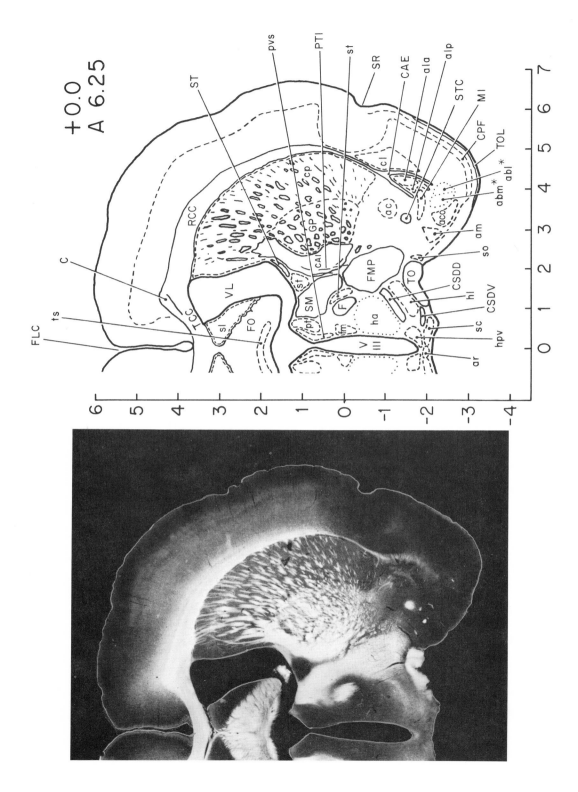

+0.0
A 6.25

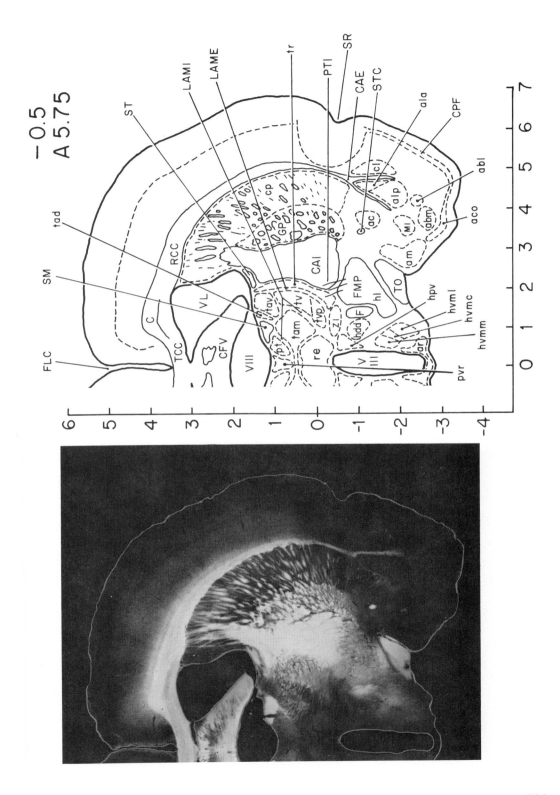

−0.5
A 5.75

FLC SM ST LAMI LAME tr PTl SR CAE STC ala CPF

TCC C tad RCC cp GPl CAl ac cl alp abl aco

CFV VL RCC av tv FMP hl TO abm am MI

VIII pt tam td ZI F hl hpv hvml hvmc

re V III ar hvmm pvr

7 6 5 4 3 2 1 0

6 5 4 3 2 1 0 −1 −2 −3 −4

221

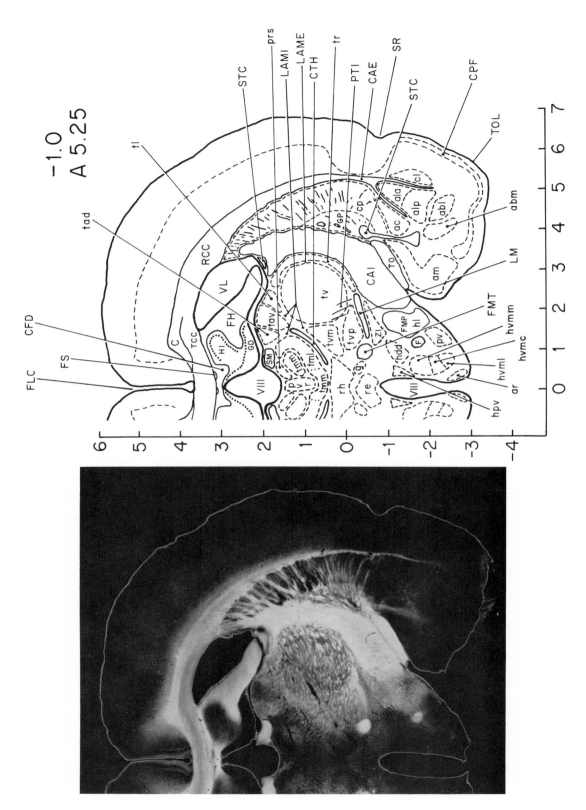

-1.0
A 5.25

222

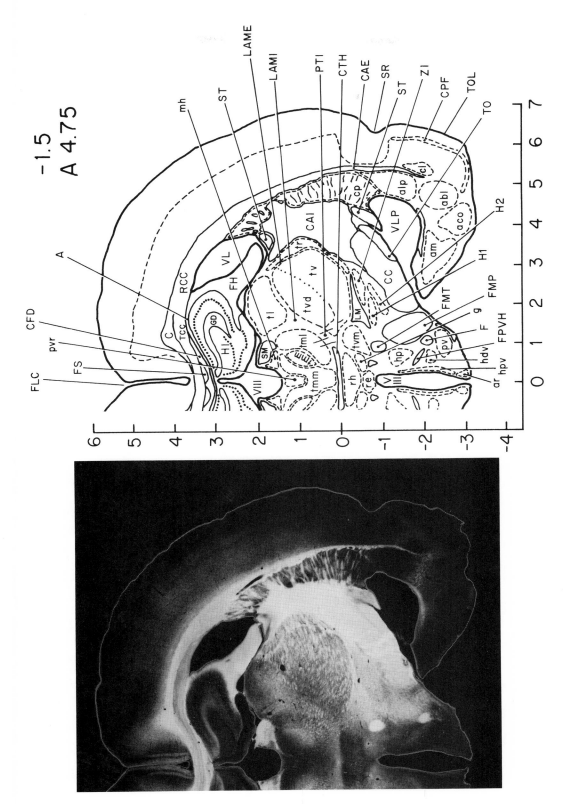

-1.5
A 4.75

223

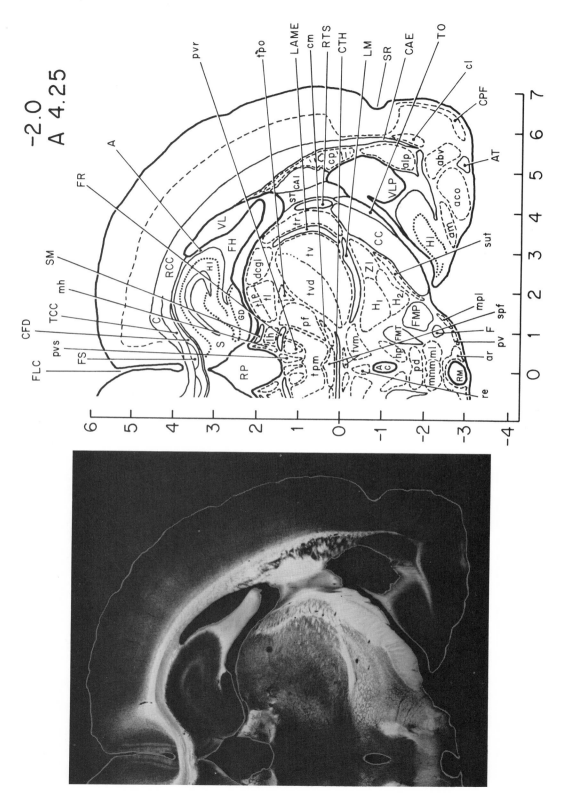

-2.0
A 4.25

224

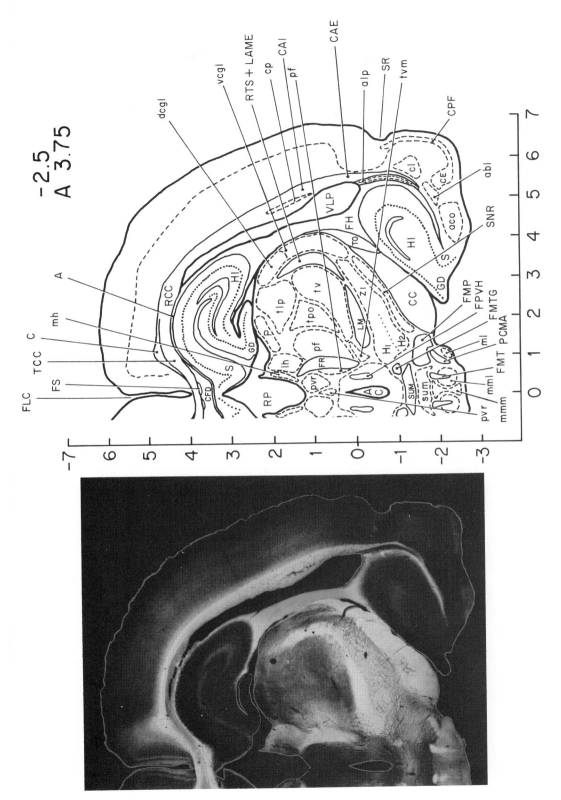

-2.5
A 3.75

225

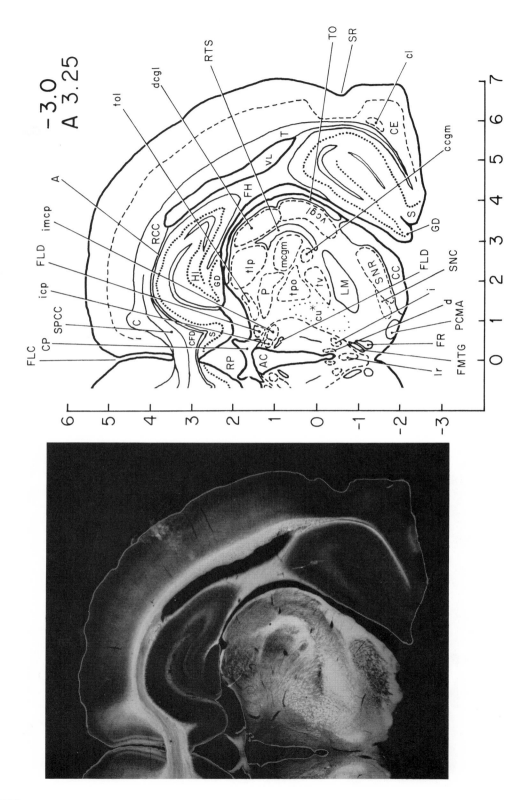

-3.0
A 3.25

226

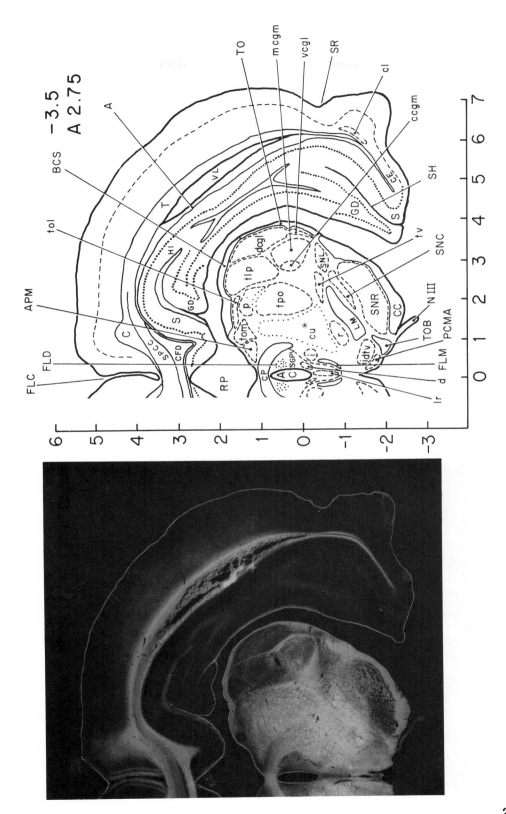

-3.5
A 2.75

227

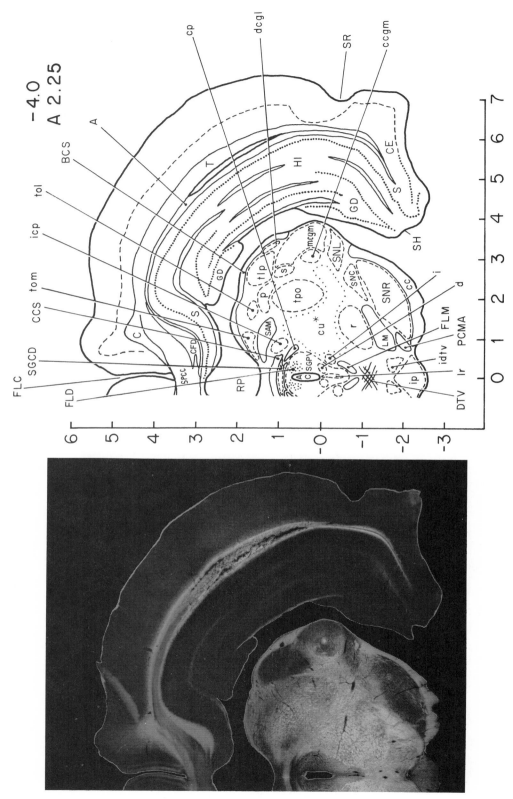

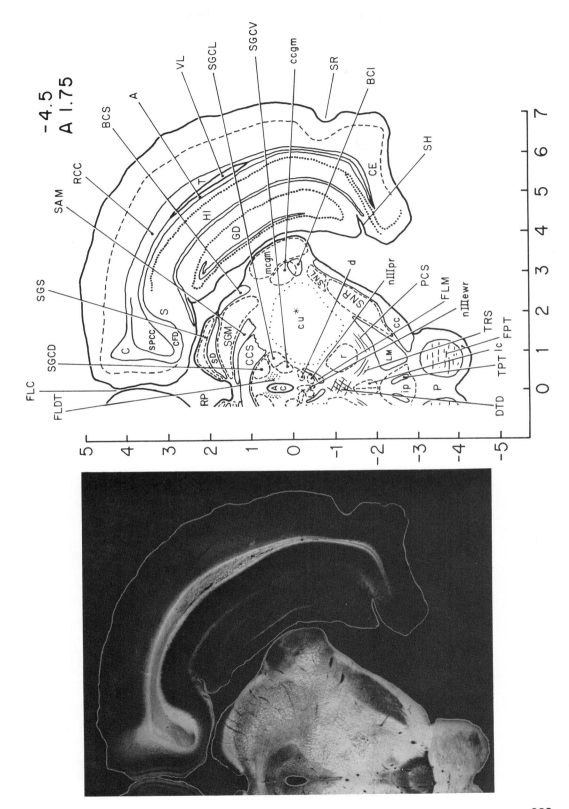

-4.5
A 1.75

229

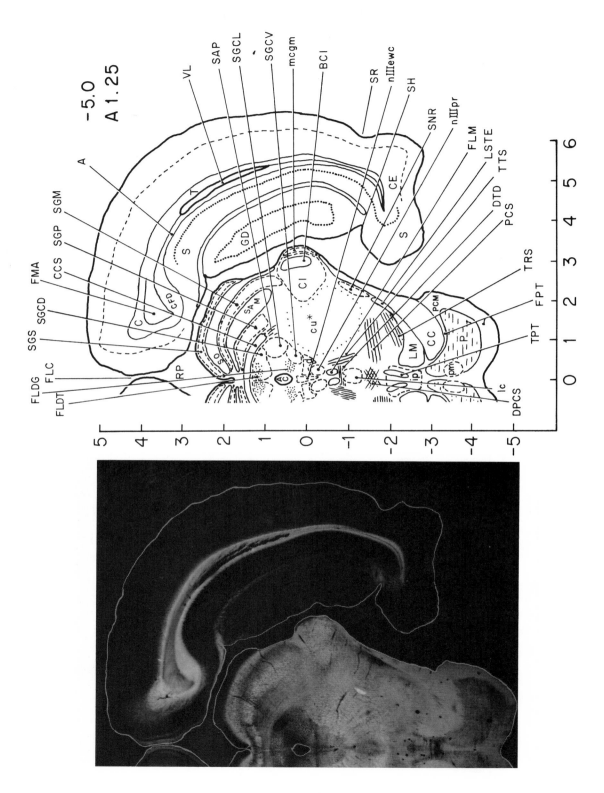

-5.0
A 1.25

VL
SAP
SGCL
SGCV
mcgm
BCI
SR
nIIIewc
SH
SNR
nIIIpr
FLM
LSTE
DTD   TTS
PCS

A

SGM
SGP
CCS
FMA
SGCD
SGS
FLC
FLDG
FLDT
RP

T
S
GD
S
CE
S
CI
SAM
C
CFD
So
cu *
c
p
PCM
LM
CC
P
pm
p
lc

TRS
FPT
TPT
DPCS

6 5 4 3 2 1 0

5 4 3 2 1 0 -1 -2 -3 -4 -5

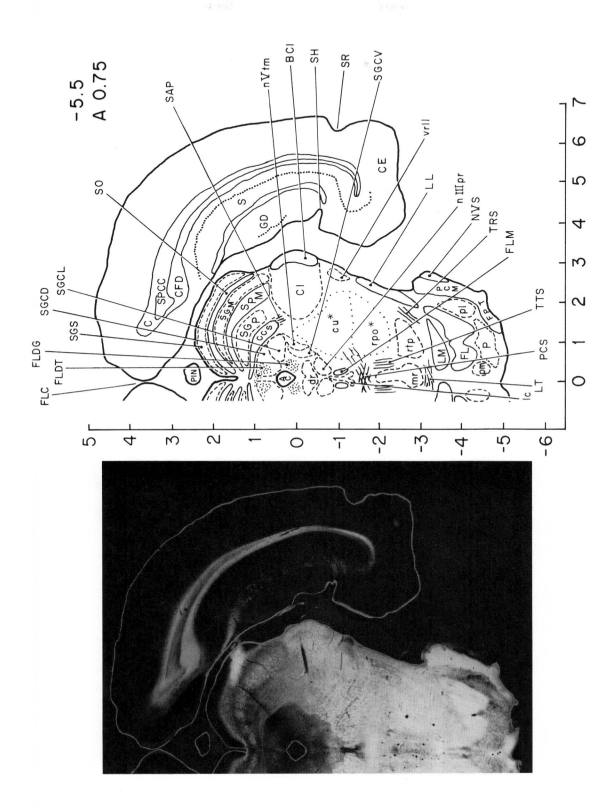

-5.5
A 0.75

231

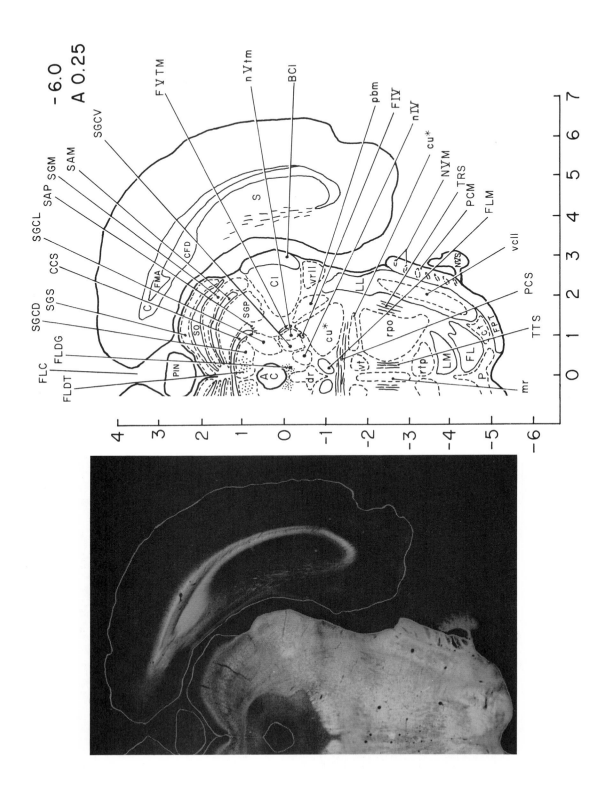

-6.0
A 0.25

232

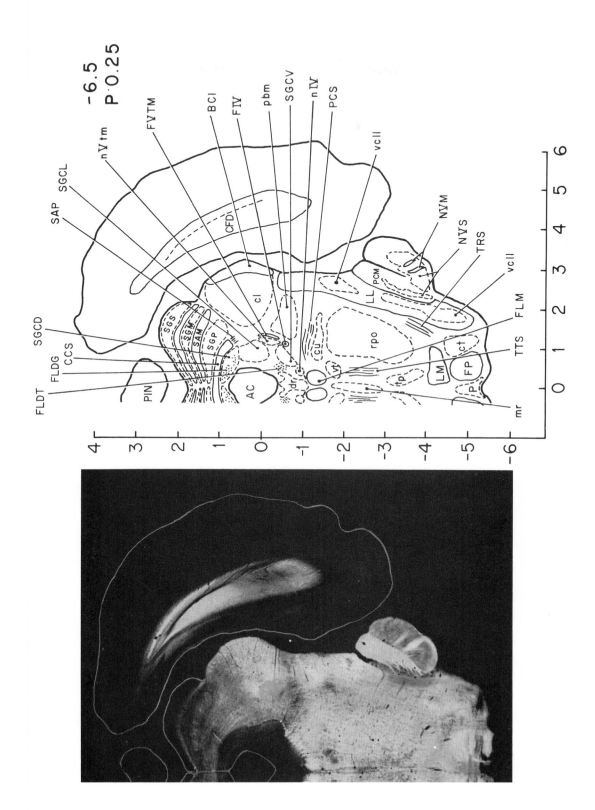

-6.5
P 0.25

233

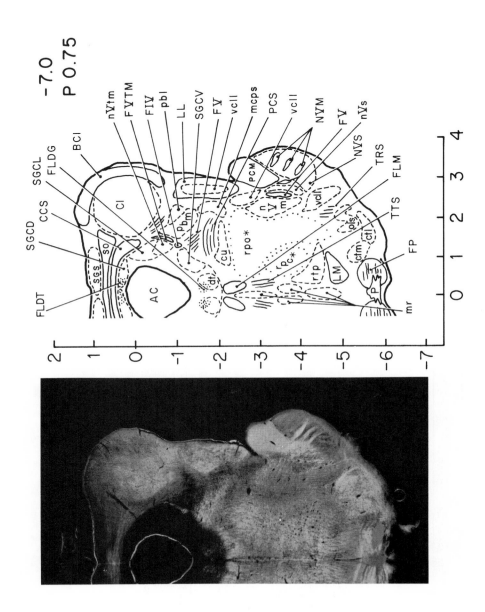

-7.0
P 0.75

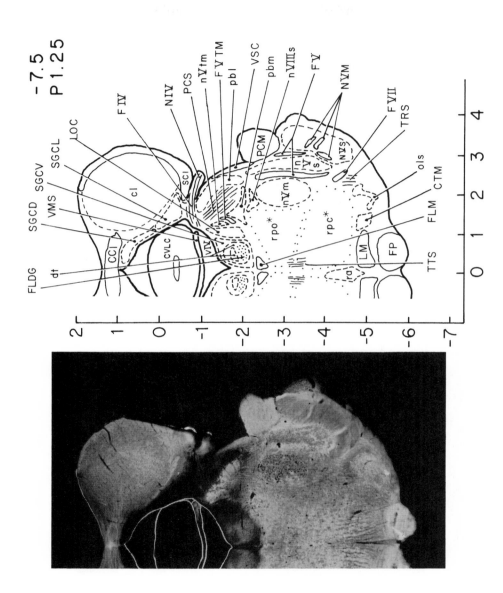

-7.5
P 1.25

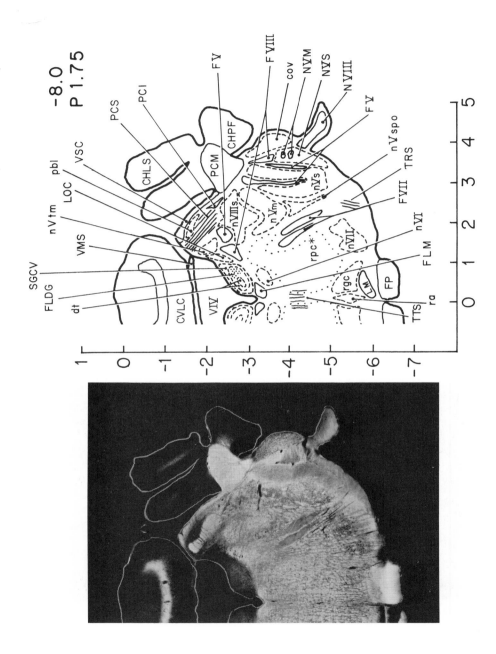

-8.0
P 1.75

FLDG
SGCV
dt
VMS
nVtm
LOC
pbl
VSC
PCS
PCI
FV
CHLS
PCM
CHPF
FVIII
cov
NVM
NVS
NVIII
FV
nVspo
TRS
FVII
nVI
FLM
ra
TTS
nVIII s
nVs
nVm
nVII
gc
M
FP
rpc*
CVLC
VIV

1 0 -1 -2 -3 -4 -5 -6 -7

0 1 2 3 4 5

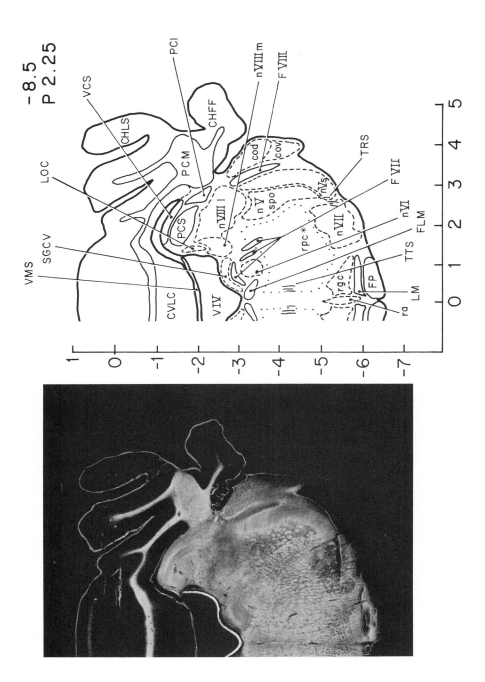

-8.5
P 2.25

237

INDEX
OF ANATOMICAL STRUCTURES
OF THE
COW AND SHEEP BRAINS

Page numbers in *italics* denote illustrations.

Alveus (hippocampus), *51, 52, 57, 58*
Amygdala, 25, *48, 49, 50*
Ansate gyrus, *43*
Ansate sulcus ("central"), 39, *43, 44, 55, 56, 57*
Anterior commissure, *48, 55, 56*
Auditory cortex, *43, 44*

Basal forebrain area, 25, *47, 55, 56*
Brachium, of inferior colliculus, *53*

Caudate, 25, *47, 48, 49, 50, 56, 57, 58*
   head, *46, 47, 56, 58*
   neck, *48*
   tail, *49, 50*
Cerebral peduncle, 25, *42, 50, 51, 52, 57, 58*
Choroid plexus, *49*
Cingulate gyrus, *46, 47, 48, 49, 50, 51, 52, 55*
Cingulate sulcus, *40, 56*
Cingulum, *46, 47, 48, 49, 50, 51, 52, 56, 57, 58*
Claustrum, 25, *48, 49, 50*
Coronal gyrus, *43*
Coronal sulcus, *43, 44*
Corpora quadrigemina, *53*
Corpus callosum, *43, 46, 47, 48, 49, 50, 51, 56, 57, 58*
   genu of, *46, 55*
   body of, *55*
   splenium of, *55*

Dental gyrus, of hippocampus, *51*
Diagonal band of Broca, *46, 47*

Entolateral gyrus, *43*
Entorhinal cortex, *42*
External capsule, *46, 48, 49, 50*
Extreme capsule, *47, 48, 49*

Fimbria (fornix), *50*
Fornix, *48, 49, 50, 51, 52*
   descending tract of, *49, 50, 55, 57, 58*
   dorsal, *57, 58*
   precommissural, *48, 50, 51, 52, 56*
   ventral, *56*

Frontal association cortex, *43*
Frontal lobe, *44*
Frontal pole, *43, 54*

Globus pallidus, 25, *48, 49, 50*
   pallidum + putamen = lenticular nucleus, *48, 49*

Habenula, *50, 56*
Habenular commisure, *55*
Habenulopeduncular tract (tractus retroflexus), *56*
Hippocampal commissure (psalterium), *56, 57*
Hippocampal gyrus, *42*
Hippocampus, 25, *50, 51, 52, 56, 57, 58*
   caudal pole of, *52*
   dentate gyrus of, *51*
   dorsal, *50, 51, 56, 57, 58*
   ventral, *51*
Hypothalamus, 25, *49, 56*

Inferior colliculus, *52, 53, 57, 58*
Inferior thalamic peduncle, 32, *57, 58*
Insula, *48, 49*
Internal capsule, *46, 47, 48, 49, 50, 57, 58*
   anterior limb of, *46, 57, 58*
Internal medullary lamina, of thalamus, *50*
Interpeduncular fossa, *55, 56, 57, 58*

Lateral geniculate body, *51, 53*
Lateral hypothalamus, *57*
Lateral gyrus, *43*
Lateral olfactory tract, *42*
Lateral ventricle, *46, 47, 48, 49, 50, 51, 52, 56*
   inferior horn of, *51, 52*
Lenticular nucleus, *48, 49*

Mammillary bodies, *42, 50, 55, 56*
Mammilotegmental tract, *56*
Mammilothalamic tract, *49, 50, 56*
Massa intermedia, *55*
Medial forebrain bundle, *57*

239

Medial geniculate body, *51, 53*
Medial thalamus, 29, *56*
Mesencephalic reticular formation, 32, 33, *52, 56, 57, 58*

Occipital lobe, *44*
Occipital pole, *43*
Olfactory bulb, *42*
Olfactory tubercule, *48*
Optic chiasma, *48, 55*
Optic nerve, *42*
Optic tract, *49, 56, 57, 58*
Orbitofrontal cortex, *42*

Parietal association cortex, *43*
Parietal lobe, *44*
Periaqueductal gray, *52*
Pineal body, *55*
Pons, 25, *42, 55, 56, 57, 58*
Pontine reticular formation, *56, 57, 58*
Posterior commissure, *55*
Preoptic area, *48*
Prepyriform cortex, *42*
Pretectal region, *56, 58*
Psalterium, *56, 57*
Pseudosylvian sulcus ("central"), *40, 44*
Putamen, *46, 47, 48, 49, 57, 58*
    caudate + putamen = striatum, *46*

Red nucleus, *52*
Reticular formation (pontine and mesence-phalic), *55*
Rhinal sulcus (fissure), *40, 42*

Sensory-motor cortex, *43, 44*
Septal nucleus, *47, 48*

Septal region, *56*
Septum pellucidum, *47, 48, 55*
Stria medullaris thalamicus, *50, 55, 56*
Stria terminalis, *49, 50*
Striatum (caudate, putamen), *57*
Subcallosal gyrus, *46, 55, 56*
Subiculum (hippocampus), *51*
Substantia nigra, *51, 52*
Substriatal gray matter, *48, 50*
Subthalamic area, 25, *51, 56, 57*
Superior colliculus, *52, 53, 56, 57*
Suprasylvian gyrus, *43*
Suprasylvian sulcus, *40, 43, 44*
Sylvian aqueduct (IV ventricle), *52, 55*
Sylvian sulcus, *40*

Tapedum, *51, 52*
Tectum (superior colliculus and inferior col-liculus), 25, *55*
Thalamocortical radiations, *57, 58*
Thalamus, 25, 29, *49, 50, 51, 56, 57, 58*
    anterior, *49*
    lateral, *49, 50, 58*
    medial, *50*
    posterior, *51*
    subthalamic area of, *51, 56, 57*
    ventral, *49*
Tractus retroflexus, *56*
Tuber cinereum, *42*

Uncus, *42*

Ventricle, III, *55*
    IV, *55*
    lateral, *57, 58*
Visual cortex, *43, 44*

SUBJECT INDEX

Page numbers in *italics* denote illustrations.

Ablation, 169, *171*
Action potential(s), 1, 6, 7
    antidromic, 12, *13*
    orthodromic, 12, *13*
Active transport, 6
Afterpotential, *10*, 11
    depolarized, *10*, 11
    hyperpolarized, *10*, 11
Agnosia, 30
Amygdala, 25
Antidromic action potential, 12, *13*
Aphasia, 30
Apraxia, 30
Artificial current channel, 19, *20*, 21
Aspirator, *171*, 183
Association systems, 26, 29. See also *Brain systems.*
    apraxias, agnosias, and aphasias, 30
    cortical association regions (multisensory), 29
    intrinsic thalamic nuclei, 29
    learning, memory, and conditioning in perception, 29, 30
Axon, 1, 2
    bifurcated, 1, 2
    collateralized, 1, 2
    limited projection, 1, 2
    specificity of growth, 1, 2
Axon hillock, 10. See also *Initial segment.*

Baillarger and Dareste, rule of, 39
Basal forebrain area, 25
Basal ganglia, 25
    amygdala, 25
    basal forebrain area, 25
    caudate, 25
    claustrum, 25
    globus pallidus, 25
    hippocampus, 25
    putamen, 25
Basis pedunculi (cerebral peduncles), 25
Binding sites, induced change of, 7
Blockade of neural structures, 169
Brain systems, 25, 26, 59, 63, 67
    anatomical, physiological, and behavioral correlates, 25
    association systems, 26, 29
    emotion systems, 26, 34
    nonspecific systems, 26, 31
    specific systems, 26, 27

Calibration, for electrolytic lesions, *172*
    for standard cryogenic systems, *190*
Cannula-guided electrode, *164*, *166*, 179
Capacitance, 150 ff., *151*
Caudate nucleus, 25

Central brain center theory of emotion, 36
Cerebellum, 25
Cerebrum, 25
    mesencephalon, 25
    diencephalon, 25
    telencephalon, 25
Chemoprobe, *178*, 191
Chemostimulation, 178 ff., *178*
Claustrum, 25
Common-mode rejection, 160 ff., *161*
Concentric electrode, *164*, 180
Coronal sections, 40, *46–52*
Cortex, 25
Cortical electrodes, *167*, *168*
Cortical projection zone, 27
Cortical surface, 39
    dorsal view, *43*
    lateral view, *44*
    mesial view, *55*
    ventral view, *42*
Cow brain. See Dissection guide.
Critical firing level, 10, *10*
Cryogenic blockade, 173 ff., *174*, *175*, *176*, *177*
Cryogenic systems, mechanisms of, *176*, 188
    simple type, *177*, 189
Cryoplate, *175*, 187
Cryoprobe, *174*, 185
Current, 145, 147
Current channels, 6
    artificial, 19, *20*, 21

Dale's hypothesis, 8
Delayed response task, 34
Dendrite, 2, *2*, 3
    spine of, *3*
Desynchronization, of EEG, *16*, 17, *17*
Diencephalon, 25
    hypothalamus, 25
    subthalamus, 25
    thalamus, 25
Differential amplifier, 160, *161*
Dipole, *18*, 19, *20*, 21
Discrimination reversal task, 34
Dissection guide (of the cow and sheep brains), 39 ff., *42–58*
    coronal sections, 40 ff., *46–52*
    cortical surface, 39 ff., *42*, *43*, *44*, *55*
    preparation of brain, for dissection, 39
    sagittal sections, 41 ff., *55–57*

EEG, 14, 15, *16*, *17*. See also *Desynchronization* and *Slow wave synchrony.*
Electrodes, 163 ff., *164*, *165*, *166*, *167*, *168*, *170*
Electrode interface (capacitance and resistance), *154*

Electrogenesis of the EEG, 14, 17, *18*
    dendritic waves, 14
    postsynaptic potentials, 14
    return currents in extracellular space, 17, *18*
    21
    small slow wave generators on cell membrane,
      14–17, *18*
    sensory evoked potentials, *20*, 21
Electromotive force, 146
Emotion, theories of, 34 ff. See also *Hypotha-
    lamic effector system.*
    central brain center theory, 36
    James-Lange's theory, 35
    MacLean's theory, 35
Emotion systems, 34 ff. See also *Brain systems,
    Limbic system,* and *Hypothalamic effector
    system.*
Equilibrium potential, 9, *10*
Equipotential lines, *18*, 19
Equipotentiality, 4
Etched metal microelectrode, *168*, 182
Evoked potentials, *20*, 21, 33. See also *Electro-
    genesis of the EEG.*
Excitability of a cell, 2, 4, 6
    critical firing level, 10, *10*
    trigger zone, 10, 12, *13*
      initial segment, 10, *10*
      dendritic, 12, *13*
Extracellular space, 3, 5, 17, *18*, 19, 21
    and impedance changes, 5
    and return currents, 17, *18*, 21
    effect of calcium on, 6
    weak currents in, 5
Extrinsic thalamic nuclei, 29

Fast prepotential, 12, *13*
F.E.T., 160
Fourier analysis of wave forms, 153, *155*
Fourier's theorem, 153
Frequency response, 153 ff., *156*, *157*
Frontal granular cortex (orbital cortex), 33

Gain and noise, 161
Glia cells, 5
Globus pallidus, 25
Gyrencephalic cortex, 39

Heat-lesion probe, *172*, 184
Hippocampus, 25
Histological techniques for examination of the
    brain, 120 ff.
    adjusting section on slide, 135, *135*
    application of cover glass, **1**, 141, *141*
      2, 142, *142*
      3, 143, *143*
    blocking the brain, 126, *126*
    blue-dot staining technique, 122, *122*
    body tissue stain, *140*
    brain, blocking of, 126, *126*
      extraction of, **1**, 123, *123*
        2, 123, *123*
        3, 124, *124*
    cardiac perfusion, 120 ff., *121*, *121*
    cover glass, application of, **1**, 141, *141*
      2, 142, *142*
      3, 143, *143*
    cutting the sections, 130, *130*
    electrodes, removal of, 125, *125*
    fiber tract stain, *138*
    floating cut section in water, 132, *132*
    freezing tissue block to freezing unit, **1**, 128,
      128
      2, 129, *129*
    gelatin embedding, *133*

    holding the knife, 131, *131*
    knife, hand position for, 131, *131*
    mounting sections on microscope slides, 134,
      134
    nucleus stain, *137*
    pedestal freezing microtome, 127, *127*
    photographic histological technique, 136, *136*
    preparations of solutions, for Rucker-Koithan
      staining procedure, *138*
      for Weil stain, *139*
    removing electrodes, 125, *125*
    sections, adjustment of, on microscope slide,
      135, *135*
      cutting of, 130, *130*
      floating of, 132, *132*
      mounting of, on microscope slides, 134
    solutions, for Rucker-Koithan staining pro-
      cedure, preparation of, *138*
      for Weil stain, preparation of, *139*
    stain, body tissue, *140*
      nucleus, *137*
      fiber tract, *138*
      Weil, preparation of solutions for, *139*
    staining procedure, Rucker-Koithan, prepara-
      tion of solutions for, *138*
Homeostasis, 34
Homeostatic functions, 25
Homestatic-like mechanisms, 34. See also
    Visceral brain.
Hypothalamic effector system, 36
Hypothalamus, 25

Impedance, 152
Implant devices, 179 ff.
    aspirator (for ablations), *183*
    chemoprobe, *191*
    cryogenic system, *188*
    cryogenic system (simple type), *189*, 190
    cryoplate, *187*
    cryoprobe, *185*
    electrodes (macro)
      cannula-guided, *179*
      concentric, *180*
      monopolar cortical, stainless steel screw, *181*
      parallel, *180*
      transcortical, platinized platinum, *181*
      twisted, *179*
    electrodes (micro)
      etched metal, *182*
      micropipette, *183*
    heat-lesion probe, *184*
    materials (list of distributors), 192
Indifferent electrode, *18*, 19
Inferior thalamic peduncle, 32
Information, 1 ff., 4. See also *Learning and
    memory.*
    all-or-none, 1
    carried by "energies", 1, *4*
    storage of, in anatomical growth, 5
      in biochemical compounds, 5
      in extracellular space, 5
      in reverberating circuits, 5
Initial segment, 10, *10*
Input resistance, 158 ff.
"Interpretative" perceptions, 30
Intrinsic thalamic nuclei, 29
IS inflection, 12, *13*

James-Lange theory of emotion, 35

Learning and memory, 5 ff. See also *Informa-
    tion.*

Lesions, 169 ff.
Limbic system, 37

MacLean's theory of emotion, 35
Mass action, 4
Medial thalamic nuclear group, 29
Medulla, 25
Membrane of the neuron, 6 ff.
Mesencephalic reticular formation, 32, 33
Mesencephalon, 25
  basis pedunculi, 25
  tectum, 25
  tegmentum, 25
Microelectrodes (intracellular and extracellular), 167 ff., *167, 168, 170*
Micropipette, *170,* 183
Motoneuron, 8, *10*
Motor system, 27
  brachium conjunctivum, 27
  cerebral peduncles, 27
  cutaneous sensory cortex, 27
  motor cortex, 27
  propriospinal system, 27
  pyramidal tracts, 27
  ventral root motor neurons, 27
  ventralis lateralis, 27
Multiple inflections, 13, *13*

Nodes of Ranvier, 7, *10.* See also *Saltatory conduction.*
Nonspecific systems, 31 ff. See also *Brain systems.*
  conscious perception (attention), control of, 31
  EEG desynchronization during attention, 31
  electrocortical excitability, differential control
    of, 33
  frontal granular cortex and the delayed re-
    sponse and discrimination tasks, 33, 34
  inferior thalamic peduncle, 32
  medial thalamocortical system, 32
  mesencephalic reticular formation, 32, 33
  nonspecific thalamocortical system, 32
  opposite actions of, 32, 33
  orbital cortex, 32
  sensory evoked potentials, enhancement of, 33
  versus specific (multisensory) systems, 31
Nonspecific thalamocortical system, 32. See
  also *Nonspecific systems.*

Ohm's law, 145 ff., *147*
  complex circuits, 150, *150*
  electrical model, *148*
  hydraulic model, *146*
  parallel circuits, 150, *150*
  voltage drop, *149*
Orthodromic action potential, 12, *13*
Oscilloscope, *159*

Papez' circuit, 37
Parallel electrode, *164,* 180
Peripheral pathway, 27
Pons, 25
Postsynaptic potential, 2, 8 ff., *10, 13*
  critical firing level, 10, *10*
  excitatory postsynaptic potential (EPSP), 9, *10*
  fast prepotential, 12, *13*
  inhibitory postsynaptic potential (IPSP), 9, *10*
  IS inflection, 12, *13*
  multiple inflections, 13, *13*
  spatial summation, 9
  spontaneous fluctuation, 12, *13*
  temporal summation, 9

Power, 159
Presynaptic inhibition, *10,* 11

Reactance, 152
Refractory phenomenon, 7
Remote excitation, 12, *13*
Remote inhibition, *10,* 11
Resistance, 145, 152
Resistivity, 147
Return currents (in extracellular space), 17, 18.
  See also *Electrogenesis of the EEG* and *Ex-
  tracellular space.*
RNA (ribonucleic acid), 5

Sagittal sections, 41, 55–57
Saltatory conduction, 8
Screw electrode, *167,* 181
Sensorimotor system, 27
Sheep brain. See Dissection guide.
Slow wave synchrony, 15, *16*
  correlation between extracellular spikes and
    EEG during, 17
  correlation between intracellular fluctuations
    and extracellular spikes during, 15
  in phase relationship of generators, 15
Sodium-potassium pump, 6
Spatial summation, 9
Specific nerve energies, theory of, 27
Specific systems, 26, 27 ff. See also *Brain systems.*
  doctrine of specific nerve energies, 27
  innate, *a priori,* perception, 28, 29
  learned, *a posteriori,* perception, 28, 29
  peripheral pathway, 27
  primary cortical projection zone, 27
  thalamic relay nucleus, 27
Specificity of growth, 1, 2
Spontaneous fluctuations, 12, *13*
Stereotaxic atlas of the rat brain, 195 ff., *199,
  202, 212–237*
Stereotaxic surgery, 88 ff.
  alignment of electrode, 107, *107*
  anesthetic dosage, 88, *88*
  antibiotic treatment, 118, *118*
  attaching animal's cable to connector, 119,
    119
  attaching skull screws, 106, *106*
  cementing connector base to skull screws, **1,**
    113, *113*
    2, 114, *114*
    3, 115, *115*
    4, 116, *116*
  cranial sutures, 101, *101*
  drilling trephine holes, 103, *103*
  fast method of anchoring electrode wires to
    skull, 110, *110*
  implanting electrode, 108, *108*
  inserting electrode connectors into connector
    base, 111, *111*
  intraperitoneal injection, 91, *91*
  length of incision, 99, *99*
  locking contacts in connector base, 112, *112*
  marking and checking coordinate points on
    skull, 102, *102*
  mounting the animal in the stereotaxic frame,
    1, 95, *95*
    2, 96, *96*
    3, 97, *97*
    4, 98, *98*
  pentobarbital dosages, *89, 90*
  placement of screw holes, 104, *104*
  postoperative recovery, 117, *117*
  removing electrode from stereotaxic instru-
    ment, 109, *109*
  scalp incision, 99, *99*
  scraping the periosteum, 100, *100*

Stereotaxic surgery (*Continued*)
  shaving the head, 94, *94*
  slicing dura mater, 105, *105*
  supplemental ether anesthetic, 92, 93, *93*
  tail-pinch test, 92, *92*
Subcortical electrodes, 163 ff., *164, 165, 166*
Subsynaptic membrane, *3*
Subthalamus, 25
Synapse, 2, *3, 4*
  electrochemical, 2
  tight junction, 2
Synchronization of the EEG. See *Slow wave synchrony.*
Systems, 25, *59, 63, 67.* See also names of systems.

Tectum, 25
  inferior colliculus, 25
  superior colliculus, 25
Tegmentum, 25
  reticular formation, 25
Telencephalon, 25
  cortex, 25
  basal ganglia, 25
Temporal summation, 9

Tertiary olfactory system, 37
Thalamic relay nucleus, 27
Thalamus, 25
  extrinsic nuclei, 29
  intrinsic nuclei, 29
  medial thalamic nuclei, 29
Time constant, 152
Transcortical electrode, *168*, 181
Transmembrane potential, 6, *10, 13.* See also *Action potential, Active transport, Postsynaptic potential,* and *Sodium potassium pump.*
  depolarized, 8, *10*
  equilibrium potential, 9
  hyperpolarized, 8, *10*
Trigger zones, *10*, 12, *13*
  initial segment, 10, *10*
  dendritic, 12, *13*
Twisted electrode, *164, 165,* 179

Vesicles, synaptic, 2, *3*
Visceral brain, 35
Voluntary motor system, 27. See also *Sensorimotor system,* and *Motor system.*
VR cell, 8, *10*